Bunjod Mamarahimov
Shuhrat Otazhonov
Shadman Namazov

Issues of primary seed production of cotton

Bunjod Mamarahimov
Shuhrat Otazhonov
Shadman Namazov

Issues of primary seed production of cotton

Improvement of seed production

ScienciaScripts

Imprint

Cover image: www.ingimage.com

This book is a translation from the original published under ISBN 978-620-8-17001-1.

Publisher:
Sciencia Scripts
is a trademark of
Dodo Books Indian Ocean Ltd. and OmniScriptum S.R.L publishing group

120 High Road, East Finchley, London, N2 9ED, United Kingdom
Str. Armeneasca 28/1, office 1, Chisinau MD-2012, Republic of Moldova, Europe
Printed at: see last page
ISBN: 978-620-8-27913-4

Mamarakhimov Bunyod Ikramovich
Otazhonov Shukhrat Ibrayimjonovich Otazhonov
Namazov Shadman Ergashovich Namazov
Malokhat Babamuradovna Khalikova

ISSUES OF PRIMARY COTTON SEED PRODUCTION

MINISTRY OF AGRICULTURE
REPUBLIC OF UZBEKISTAN
NATIONAL CENTER FOR KNOWLEDGE AND INNOVATION
AGRICULTURALLY

Mamarakhimov B.I., Otazhonov Sh.I., Namazov Sh.E., Khalikova M.B.

ISSUES OF PRIMARY COTTON SEED PRODUCTION

UDC: **633.51:575:631.52**

The monograph is devoted to the issues of improvement of seed production methods applied in the farms of the republic on the basis of comparative analysis of scientific and practical experiments on elite seed production conducted on cotton varieties. It involves the study of methods of cultivation of super-elite and elite seeds used in the world practice and the development of convenient methods of cultivation of original and super-elite cotton seeds on their basis, the influence of self-pollination method on the preservation of economic characteristics, technological indicators of fiber quality, productivity and yield of varieties.

The monograph can be used by seed breeders and seed scientists, plant breeders, research scientists, undergraduate and graduate students.

INTRODUCTION

The most important condition for increasing the quantity and improving the quality of cotton production is the improvement of production and preparation of seeds, which have passed the State test, patent-protected registered varieties. The main problems here include the following: poor integration of breeding, variety testing, seed production and seed quality control; insufficient investment in the seed production system; lack of a coherent system of variety identification and seed quality assurance at all stages - from breeding to seed consumer, insufficient training of personnel to work in the new market conditions.

To date, it remains problematic to implement a modern market scheme for production and processing of super-elite, elite and reproductive seeds, as well as to re-establish a system of seed quality control that meets the requirements of international norms and rules.

On this basis, the question of the necessity to study actual problems of cotton seed production development, to carry out detailed scientific developments aimed at improving seed production work in the conditions of transition of seed production to private ownership in farms has been put before us.

Fulfillment of these tasks depends to a great extent on the well-established work of elite-seed farms, which are entrusted with the production of the necessary amount of super-elite, elite seeds, preservation of the typicality of the breeding variety, its economically valuable traits and technological properties of fiber.

Analytical literature review of scientific researches of foreign scientists carried out in the field of breeding and seed production of agricultural crops shows the effectiveness of different methods of production of super-elite and elite seeds.

Scientists of such leading cotton-growing countries as the USA, China, India, Brazil, Australia, etc. are conducting intensive research on the production of genetically aligned cotton varieties, inheritance patterns, variability, form-formation processes and correlation of traits for use in practical seed production.

As a result of research of foreign scientists on application of various methods of elite cotton seed production their advantages and some existing disadvantages were revealed. The efficiency of the methods used to create genetic homogeneous sown varieties of certain value for the seed production process has been confirmed.

In addition, they note the relevance and demand for studying some methodological aspects of improving the methods of alignment of genetically homogeneous seed material and its use in seed production. Intensive research on preservation and improvement of early maturity, high yield with high yield and fiber quality of cotton varieties is also carried out. It is believed that the most efficient and fastest method of selecting the best cotton types is individual selection combined with self-pollination and subsequent testing of the progeny. The effect of self-pollination on the viability and other properties of cotton varieties and forms has been studied by many scientists, but there is no consensus on this issue.

Research on studying and selecting methods of elite seed production is being actively conducted. Since 1985, the question of methods of elite cotton seed production has been debated. The controversy has not subsided until now. This is explained by two reasons. The first is that, unlike other crops, the breeder-originator of a cotton variety after its zoning does not take part in elite production. It is transferred into the hands of seed breeders of elite farms, while for all crops both in our country and abroad (including cotton), every year the originator of the variety himself prepares elite seeds according to the method he considers most expedient for the preservation of the variety or its improvement. Then he transfers them to special farms for multiplication.

Based on the methods used, it is not always possible to conserve varieties with high parameters of economically valuable traits.

Every year, each elite farm conducts field scraping of families, harvesting of the best plants in the form of individual selections, test samples from selected families, as well as family-by-family collections for laboratory analysis. The obtained statistical material is processed with the help of a calculator, then, variation series are compiled, average indices reflecting the quality of elite materials are determined, rejection is carried out with comparison of all signs characterizing the family and average indices of variation series of families and individual selections. The cumbersome nature of the analyses and processing of their results raises the question of improving this labor-intensive process, involving modern information methods, including the use of computer programs.

The current methodology in elite work uses outdated parameters of fiber quality assessment: such as fiber strength, staple fiber length in mm, which does not correspond to international standards, where parameters are given according to the HVI system. It is necessary to switch to the assessment of technological properties of fiber, which is valid in the main cotton-growing countries of the world and accepted in the world cotton market.

For the development of methods of evaluation and rejection of elite material, the issues related to the reduction of labor-intensive and time-consuming costs of calculating variation series, improving the quality and competitiveness of raw cotton and fiber on the world market are of primary importance. Qualitative and quantitative indicators of produced cotton fiber and other by-products largely depend on the level of selection and rejection of seed material. Modern means of office equipment (computers) allow to simplify the labor-intensive process of statistical processing of initial data on elite material of multiplied cotton variety. Individual selection, cleaning, rejection of families are carried out by differently qualified seed breeders, who can not always give a reliable assessment of the left families, are not quite able to recognize the genotype required by the author and as a result of erroneous selection and rejection of families can be a complete loss or sharp deterioration of the original economic-valuable traits, respectively, technological qualities of fiber. Therefore, it is necessary to improve the methods of production of original seeds aimed at preserving the economic-valuable features of the variety on the basis of maintaining the genetic homogeneity of produced seeds with varietal and sowing qualities and obtaining a sufficient number of seeds for varietal renewal.

The purpose of our research was to carry out a comparative analysis of the accumulated scientific and practical experience in the development of elite seed production and adapt the most acceptable methods in the emerging conditions of seed production by farms of the republic, to study the methods of production of super-elite and elite seeds used in international practice and to develop on their basis the most acceptable method of production of original and super-elite cotton seeds with the inclusion of a computer program for automation of the production of original and super-elite cotton seeds.

In order to achieve the intended goal, the following research objectives are formulated:

- simplification of the process of statistical processing of the results of laboratory analyses of elite material;

- on the basis of internationally applied methods of super elite seed production to improve individual links (elements) of seed production activities in seed farms.

- comparative evaluation of the reproduction efficiency of original and super-elite cotton seeds according to the current and proposed methodology;

- introduction in elite-seed farms of assessment of basic technological properties of fiber of trial samples and individual selections according to the international system HVI, generally accepted in the leading cotton-growing countries of the world;

- Improving the accuracy and reliability of scraping results to select the best families of the propagated variety with quality parameters according to international parameters, in order to maintain or improve the fiber quality characteristics of the variety;

- to develop computer programs for accelerated and more reliable evaluation of author's material for use in breeding and seed production work, at the final stage of formation and refinement of varieties;

- revealing the influence of self-pollination on varietal purity of elite seed material;

- identification of the optimal variant of modal selection of plants and families aimed at changing the variety structure in the desired direction in the corresponding group of plants.

The significance of the research results lies in the fact that on the basis of self-pollination the efficiency of creating purebred material of multiplied cotton varieties has been confirmed. On the basis of analysis of application of various methods of production of original cotton seeds the ability to variability and correlation dependence of main economic-valuable features of seeds of varieties propagated in different regions of the republic was revealed. The methodology for production of super-elite, elite seeds of registered and new varieties has been improved.

CHAPTER i LITERATURE REVIEW

§1.1 Primary seed production as the foundation of theoretical and practical basis for realization of variety potential

Variety is the real basis for growth, stabilization of production and improvement of crop production quality. It is closely related to natural and climatic conditions, zonal technologies, technical means, level of industry management and determines the fate of cultivated and spread of new species [65.P. 5-16, 66. P. 536, 153. P. 686, 252. P. 34-36].

The main requirements that new varieties should meet are a high degree of adaptation to the conditions of the proposed zone of their growth, specified parameters of productivity, quality, resistance to abiotic and biotic stresses, stability of yields under unstable hydrothermal conditions. New varieties should surpass those cultivated in the zone in terms of the main indicators. These provisions form the basis of the concept of variety model and determine approaches to solving the problem of optimization of the breeding process.

A promising direction is the use of computer technologies that provide information support of the breeding process from laboratory research to field experiment, allowing to promptly analyze and use subject, agrometeorological, technological, analytical and other information.

The most widespread for solving breeding and genetic, as well as production tasks, computer databases are currently being used:

- passport, containing the main parameters of the collection samples

(crop name, botanical name of the specimen, place of origin, etc.);

- descriptive, containing basic phenotypic characteristics (inherited traits, recorded visually, constant under all conditions);
- evaluative, including characteristics that depend on the impact of environmental factors (yield and other economically valuable traits, genetic, biochemical, anatomical, physiological and morphological

indicators, resistance to diseases and adverse environmental factors) [34. C. 32-40].

Modern means of office equipment (computers) allow to simplify the labor-intensive process of statistical processing of initial data on elite material of multiplied cotton variety, to avoid human errors. In order to use these means of office equipment in seed production practice of elite farms of the republic, it is planned to develop a special computer program available for each seed grower.

The work, on the basis of the developed program assumes introduction in domestic seed production practice of evaluation of technological properties of trial samples, which has not been done before, according to the international system HVI, generally accepted in the leading cotton-growing countries of the world. This will allow more accurately and reliably allocate the best seeds of multiplied variety by fiber according to international parameters and thus control, preserve and in some cases, even improve the quality characteristics of cotton fiber, the main product of cotton production [191. P. 110-112].

If a person continues to select and thus strengthen some characteristic trait, he always almost certainly modifies unintentionally other parts of the structure due to the mysterious laws of correlation [67. P. 350].

The doctrine of correlations, pushed to the background by the idea of organisms as a mosaic, again comes to the forefront as a result of the establishment of genetic linkages and a deeper understanding of the connectivity of functions and organs in the organism and the establishment of morphophysiological correlations. Thus, contradictory phenomena are dialectically combined: on the one hand, great possibilities of rearrangement of properties and traits that distinguish organisms, breeds, varieties, and, consequently, great possibilities of practical breeding, and on the other hand, undoubted facts of existence of correlations [49.P.20-40 , 50. P.262-287, 260. P.137-141].

As a product of natural and artificial selection, correlations and correlation series play an essential role in the evolutionary transformation of plants and animals. Each insignificant, seemingly insignificant, trait grows a multitude of related traits and thus becomes an element of the program of future evolution [37. P. 23-60].

If in the XIX and the first half of the XX century scientists mainly described correlations between morphological, embryological, ecological, psychophysiological, biogeographical features in the most diverse groups of creatures, then in the second half of the XX century and nowadays more and more attention is paid to the use of correlations, first of all, in breeding and systematics.[37. P.23-60].

Many examples of such use of correlations can be given. Let us limit ourselves to the following.

It was shown that selection of fodder beets by root length allows increasing the average fruit weight, but leads to a decrease in dry matter content [271. P. 66-57].

Selection for high and low fat content in maize seeds leads to reduced fertility, early maturity and resistance to low temperatures [250. P. 163-222].

Selection to increase the coarseness of buckwheat seeds leads to an increase in the yield of this crop. An example is the variety Bogatyr, created by Shatilov exactly by the method of multiple selection of the coarse fraction from a local variety of Orel province [251. P. 179].

Identification of a close relationship between the weight of ears per plant, as well as between the weight of the main ear and grain productivity made it possible to abandon selection for grain productivity of grain ear crops and to select for the total weight of ears or the weight of the main ear. Such selection makes it possible to cull all unproductive material from plants selected in the field even before threshing [158. P.373, 159. P. 284,165 P.162]. There are reports that it is possible to distinguish new types of plants at the seedling stage, based on a significant correlation between the traits of young and adult plants [240. P.130-132].

Unfavorable factors in the form of moisture deficit, strong

salinity, insufficient nutrition, accumulation of pathogenic soil microflora and some pests have always restrained the growth of productivity of grain crops when they are placed on the predecessors. About 60% of spring barley crops in Kyrgyzstan are concentrated in rainfed farming zones, where the yield is fully dependent on natural moisture. The greatest amount of precipitation in these zones falls in winter and early spring periods, while in the summer months it is scarce and does not have a significant impact on the formation of yield. The negative impact of drought is intensified by the action of strong drying winds, to which large areas of rainfed lands in the Chui Valley are exposed [39.]. All components of the chemical composition of grain vary significantly depending on the genetic properties of varieties and growing conditions [98. P.57-58].

One of the main indicators of barley grain quality and indicators of its fodder and brewing properties is the chemical composition of the grain. Nutritional value depends on the content of protein and starch in the grain. Other substances in the grain are less, and they have less nutritional value [204. P.6-9, 218.].

The system of integrated plant protection, although it allows to increase the level of realization of genetic potential of a variety, acts as an important but schematic set of techniques and methods of influence on plants and external agro-ecological environment without deep evolutionary-genetic and ecological-genetic substantiation of components of the whole system of adaptive resource- and energy-saving and environmentally safe agricultural production [88. C. 432, 89. P. 767, 158.P. 373, 159.P. 284, 177.P. 121-123, 184.P. 116-117, 266.P. 594, 270. C. 336]. It is necessary, from our point of view [162. P. 43-48,163. P. 31, 168. P. 364 , 169. P. 301], to deepen the general biological and genetic concept of management of adaptability of variety genotype in the process of its cultivation, i.e. in seed production, crop production and farming.

The genetic interpretation of trait correlations is now reduced mainly to the linkage of genes and their pleiotropic effect, when a change in one or more genes can cause a change in a number of traits. Polygenes controlling quantitative traits may be in numerous linkage groups and very often in reciprocal relationships, which reduces the frequency of recombinants in splitting populations. In this regard, selection for individual traits usually leads to the deterioration of the complex of traits of the population, both in breeding and seed production practice [16. P. 42-45, 17. P. 78,18. P. 40].

In cotton, the majority of economically valuable and morphological traits are paired in a certain way. Researchers have different opinions about the degree of trait pairing, but these opinions are mostly subjective.

Correlations of quantitative traits in cotton were once given great importance [84. P.233-235]. Indeed, the characterization of the first domestic varieties of the 20s reveals strong inverse relationships between such leading traits as early maturity and raw cotton yield, length and fiber yield. An inverse relationship was found between early maturity and cotton yield [10. P.9-10, 13. P.23, 18.P.4, 69. P.282-287, 70. P.9].

Many foreign and domestic researchers have studied the interrelationship of cotton traits, mainly in the direction of identifying the conjugation of yield and its elements with other economically valuable traits. Yield is positively correlated with boll weight and their number [7. P.12-13] and negatively with fiber length [8. P.7-9].

A high positive correlation of raw cotton yield with fiber index and fiber weight in the boll is observed. Fiber yield is inversely correlated with length [15. P. 20, 77. P. 80-89].

A negative correlation was found between fiber length and yield, and a positive correlation between fiber length and fiber yield [59. P. 65-71]. It is estimated that with an increase in fiber length by 1 mm, the yield decreases by 4%.

Negative relationship between the length of the growing season and fiber yield is noted [20. C.19-21, 105. C.48-61 , 164. C.3-8 , 180. C.29-31, 185. C.106-108 ,240. P.130-132] and come to the conclusion that such correlation is caused by genetic peculiarities of a variety or form. At the same time, a positive correlation between the length of the growing season and fiber yield is noted [224. P.177-179, 225. P.24, 238. P.24, 239. P.19-22].

Negative relationship between the length of daylight hours and fiber yield was observed in studies on the reaction of varieties and hybrids of populations of cotton to the length of daylight hours. It was found that under conditions of 10-hour illumination in hybrids F_2 there is an intermediate character of inheritance of correlation traits on fiber yield and daylight hours with a bias towards high-water parental forms [110. P.110, 111. P.221].

A negative correlation between fiber yield and its strength is noted [58. P.29-30] .A positive correlation was found between the weight of raw cotton of one boll and fiber yield, and a negative correlation between the number of bolls and the weight of raw cotton of one boll.[69. P.282-287].

Genetic analysis cannot be complete without studying phenotypic, genotypic correlations of traits and heritability. As is known, correlations are due to the pleiotropic effect of genes or their coupling. In cotton, the majority of economically valuable traits are negatively correlated, which in turn makes it difficult to create varieties with the complex of economically valuable traits necessary for breeders. Therefore, it would be reasonable to study correlation interactions of traits. Many researchers have paid attention to phenotypic correlations, which can sometimes differ significantly from genetic ones. Genetic correlations are less well studied.

In elite-seed crops of the released cotton varieties An-Bayaut-2 and C-6524, along with obvious biological and mechanical impurities, plants deviating from the variety description only by individual morphological traits are found. By rejecting atypical plants in nurseries, complete homogeneity of morphological traits of plants is not achieved; every year it is necessary to select typical and reject atypical families and plants.

In this case, not all genetic variability of the population is used, but only some part of genotypes, and the initial heterogeneity of the population changes, reducing its competitive ability, as well as reducing the efficiency of selection of elite material in nurseries. On the other hand, biological contamination of a variety leads to its degeneration. At the same time, in rejected plants of both varieties, the average population productivity of plants also decreased naturally by years of elite reproduction [194. P.8-9].

The following is a brief review of works on trait correlations.

High positive correlation between the productivity of raw cotton of one plant and fiber strength is noted [102. P.192-194] determined a negative correlation in this pair of traits.

There is a positive paratypical correlation between the productivity of raw cotton of one plant and fiber yield, [74. P.79-80] A negative genetic correlation between the components determining the quality (length, fineness) and quantity of fiber (index, fiber yield) has been established [78. P.109, 271. P.57-66].

Interesting data on interdependence of fiber yield components are reported [45. P.467, 97. P.100-102, 130. P.109-110] It is noted that there is a medium positive correlation between the weight of 1000 pieces of seeds and fiber index, between fiber index and fiber yield there is a high positive correlation, between fiber length and seed index there is a weak negative correlation, between seed weight and fiber yield there is a weak negative correlation, between length and fiber yield there is a weak negative correlation. These correlations are relative in nature and depending on varieties and growing conditions can change from strong to weak, from positive to negative and vice versa.

The works [13. P.23, 229. P.26-34, 231. P.120-121, 247. P.22] raise the most important questions related to variability and inheritability of the length of the growing season and outline ways to influence the cotton plant in order to increase its early maturity. High quality seeds are important indicators in determining early maturity [13.

P.23, 211. P.146, 212. P.148, 214. P.24-26].

It has been established that fiber yield in cotton is a complex trait determined by the number of fibers on the seed, their absolute mass and the mass of the seed itself. Breeding in F_2 of higher yielding plants than parental forms depends on the inheritance by the offspring of high mass of fibers of one parent and a large number of them in the other, and the exclusive role of pair selection for crossing by elements determining fiber yield is noted [105. P.48-61, 118. P.115-116].

The negative correlation between fiber length and fiber yield is noted to varying degrees, from weak to strong [10. P.9-10].

Proved the possibility of creating hybrid material, and then varieties combining high fiber yield with long fiber, an example of this are varieties C - 6029, Kzyl - Rawat, C - 6524, Namangan - 77, C - 6530, C - 6532, Surhan - 5 (2001) [11. P.144, 18. P.40].

Seed breeding is one of the ways to manage variety adaptability based on the ecological-genetic principle of creation and existence of ort genotype. It should be considered as the most effective and organizationally available means of intensification, biologization and ecologization of processes in crop production on the basis of allocation of seeds with high yielding properties due to epigenetic processes of the preceding ontogenesis. In this case, the process of formation of yield properties of seeds, providing the realization of genetic potential of variety productivity in specific agro-ecological conditions, includes both adequate replacement of technogenic means with biological ones, and more rational use of natural and technogenic resources (ecological seed production), thus providing energy efficiency on the basis of exogenous impact on the genotype through the phenotype of the variety, sustainability, environmental protection and profitability of agricultural production as a whole. This can be done purely technologically in the process of cultivation of the variety and seed formation on the mother plant, as well as on the basis of selection of high-yielding batches of varietal seeds according to their yield properties when growing them both in different agrotechnical and agro-ecological conditions [33.

P.354, 107. P.6-7, 111. P.221, 117. P.22-39]. Studies on intra-varietal biotypic selection, seed fractionation, and seed reproduction have shown epigenetic conditioning of yield properties of varietal seeds. Numerous experiments, both in our country and abroad, on sowing dates, depth of seed embedding, the effect on seeds and plants of growth stimulants, various physical factors, seed cultivation in different soil and climatic zones, also testify to the ontogenetic conditioning of yield properties of varietal seeds. This calls into question the existing theoretical basis of varietal renewal in seed production [111. P.221, 162. P.31, 170. P.].

High variability of environmental conditions in time and space, impossibility to control and regulate them, causes high variability of yield and its quality. It is believed that selection of varieties with sufficiently high agroecological adaptability on the basis of deeply justified zoning within the boundaries of cultivation of a given crop contributes to reducing the amplitude of yield fluctuations [84. P.233-235, 85. P.664, 131. P.8, 201. P.8]. However, the adaptability of a variety is realized through seeds bearing the imprint of embryogenetic (epigenetic) changes of different types of environmental impact on cells, organs, organism in the process of differentiation, growth and development [161. P.97, 162. P.43-48].

Modern seed production has poorly developed scientific and theoretical base, it is mainly represented as a branch of agricultural production [60. C.40-44, 61. C.136-143, 95. C.3-5, 96. C.3-5, 120. C.129-130, 160. P.48], and this is not quite right. And since the variety acts as one of the powerful levers of scientific and technical progress in agricultural production, the theoretical basis of its use in production should be deeply scientifically substantiated. Seed breeding as a science is the foundation of theoretical and practical basis for the realization of genetic potential of variety productivity, i.e. its object is a variety and methods of preserving its genetic potential in the process of reproduction, on the basis of seeds, reflecting the specific state of the genotype of the variety, its ontogenetic adaptability, expressed in gene expression and cytoplasmic heredity, reflected through the morphophysiology of seedling organs of formed seeds.

New successes of biology in 70-80 years, which connected the problem of species with the doctrine of specificity of genetic systems in eukaryotic and prokaryotic species, showed the universality of the phenomena of mutagenesis and recombinagenesis, revealed the basics of molecular organization of genomes, the importance of macromutations, the role of not only divergence, but also various forms of fusion of plasmas of different species, gradually changed the content of a number of classical postulates of the theory of evolution and fundamentally changed the theoretical provisions of breeding. All this allows us to take a critical look at the methodological basis of modern seed breeding, seed science, and variety genotype as a product of evolution, existing on the basis of four principles formulated by us - evolutionary-genetic, ecological-genetic, and natural and artificial selection [164. P3-8., 167. P.100]. The presence of a continuously running evolutionary process is directly revealed in the phenomena of speciation, chromosomal structure of species and its organization, in the dynamics of ecological and geographical differentiation of populations, in polymorphism and individual variability, in the constant occurrence of mutations and recombinations [76. P.49, 80. P.26-27, 87. P.587, 118. P.115-116, etc.]. In all these cases, the chain of events is caused by the action of natural selection.

Epigenetic form of variability is inherent in units of heredity of different hierarchical levels: transcriptons, loci, epigenes, epigenes, superloci, chromosomes, genomes. Epigenetic heredity and variability of traits play a major role in many micro- and macroevolutionary events that are accessible to the action of natural selection [168. P.364]. All this clearly indicates that the genotype of a variety is created and exists on the basis of evolutionary and ecological-genetic mechanisms based on natural selection and largely enhanced by artificial selection of genetic variability [76. P.49, 84. P.233-235, 86. P.3-6, 87. P.587, 165. P.162, etc.].

In seed production, seed science and plant breeding we deal with epigenetic variability during ontogenesis - this is the induction of gene activity caused artificially by various agronomic practices (sowing dates, irrigation, fertilizers, biostimulants, etc.), naturally by various environmental conditions, pathogen products, etc. It leads to changes in morphogenetic and morphophysiological traits (phenes) during ontogenesis. It leads to changes in morphogenetic and morphophysiological traits (phenes) during ontogenesis. Based on this provision, a methodology for assessing the yield properties of seeds and their yield potential was developed [149. C., 155. C., 167.C.,168. C.].

1.2 History of cotton seed production and methods of original seed production

The first attempt to import American cotton seeds in 1871 was unsuccessful. Then seeds of late-ripening C-Islands (seaside), which were not adapted to the natural conditions of Central Asia, were imported, and later seeds of faster-ripening varieties - King, Russell, Cleveland, etc. were imported [1. P.127]. Until 1914 in different regions of Central Asia there were 15 plots that could produce 18-20 thousand poods of improved seeds per year, while the need for them was defined as more than 2 million poods [256. P.5]. In 1914. S.V. Poniatkovsky proposed a number of measures, which were supposed to attract some groups of owners and entire rural societies to multiply seeds of the best varieties, as well as to organize the cleaning of cotton yields of these farms at cotton mills under a special contract
[179. C.80-90].
For the multiplication of varietal cotton seeds in 1924 organized a special trust of state seed farms "semkhlopok" which in 1931 reorganized into "sovkhozkhlopok".

In 1931, the breeding and seed production work was restructured. Seed farms were liquidated, and instead of them, a network of elite-seed farms was organized in collective and state farms for annual reproduction of elite seeds in the first reproduction of cotton varieties [179. P.80-90,

197. P.83].
After World War II, major changes took place in elite cotton breeding work. Since 1951, in order to improve the hereditary basis of the variety, the method of intra-variety crosses was additionally introduced, which showed that pollination of flowers with a mixture of pollen from plants originating from different regions increased the yield of raw cotton up to 20% compared to natural pollination in some cases [19. P.11-12, 115. P.59-61]. This method was supported by [81. P.27-28, 108. P.29-31, 176. P.33-35] and others, who argued that the use of intra-sort crossing methods in elite-seed work of cotton allows to preserve and increase their vitality. The acceptance of intra-sort crossing was opposed by [12. C.26-34, 24. C.30-33, 25. C.35-36, 113. C. 9. 26- 29, 157. C.34-35, 258. P.35-38] and other specialists who claim that intra-varietal crosses in cotton seed production do not give a positive effect, but, on the contrary, create a mass of non-typical and deviated plant forms. In this regard, varieties quickly lose their original economically valuable traits and technological fiber qualities, and this is the reason for the removal of varieties from production. From the genetic point of view, the method of intra-varietal crosses widely used in cotton growing is simply harmful, as it gives heterosis (a surge of vitality) only in the first generation. In the subsequent generation, the usual Mendeleevian cleavage on multigenic and complex cleavage on quantitative traits, i.e. on traits determined by many genes, occurs. In thousands of experiments on self-pollinators intra-varietal crosses were fruitless.[264. P. 186].
As pointed out [49. P.262-287], intra-varietal crosses have a positive effect if the variety is subject to cross-pollination.

There is an opinion [17. P.78, 65. P.5-16, 275. P.89-90] that a variety is a population in which there are interrelated processes of improvement and deterioration of properties of individual individuals, resulting in systematic selection and long-term testing of offspring to maintain the variety (population).
There are dozens of methods of elite seed production in the USA and other foreign countries [51. C.12,52. C.87-90, 124. C.20-23, 199. C.8-

12, 200. C.45-46, 234. C.67, 257. C.483, 272. C.3-7].

In Uzbekistan, in the mid-80s of the last century, the method [180. P.29-31] was tested as a simplified method of elite seed production based on selection without verification by progeny.

As emphasized [101. P.5-7, 104., 106. P.4-5, 128. P.98-99] in the current Instruction on production of seeds of elite and first reproduction of cotton varieties, it is not provided for elite-seed production taking into account genetic features of varieties, the origins of their breeding and selection, duration of being in production [7. P.12-13].

In the Republic of Uzbekistan, in connection with the replacement of cotton variety mixtures with purely varietal ones, in the 30s of the last century, elite-seed breeding work was organized based on the individual-mass approach to the evaluation and selection of elite material. Usually, any refined variety in the most distant reproductions (reseeding of elite) persistently retains its economic and biological traits that determine yield. Undesirable changes in heredity for these and other traits are not group, but individual, they occur rarely, only in individual plants and therefore can not quickly, without a very long action of natural selection to affect the deterioration of the variety [10. C.9-10, 83. C.102-103, 84. C.233-235, 91. C.219, 103. C.194-196, 116. C.29-31, 136. C.64-66, 147. C.289-293, 213. C.20-21, 242. C.31].

In TNIISSTH as a result of many years of research (from 1972 to 1985) developed a new method that allows you to preserve the original qualities of the variety for a long time.[180. P.29-31]. This method was compared with the method of individual selection adopted in production, with a check on the offspring without intra-variety crossing. Experiments conducted on the variety Ashkhabad-25, show that the reproduction of cotton seeds by the proposed method does not deteriorate such indicators as boll size, yield and technological qualities of fiber. Advantageous or identical results indicate that at high reproduction rate by the new method it is possible to obtain high quality seeds with the lowest costs [180. P.29-31]. Such a method has not found application in Uzbekistan, although it was approved in

Turkmenistan. This is due to the fact that during the testing of the method in 1984 a number of shortcomings were noted and it was decided to conduct only its production test to reconsider this issue [122. P.31-33, 134. P.56-58, 145. P.22]. The essence of the new method consists in collecting a large number (based on need) of individual selections. They are cleaned, scraped by seeds and by the length of fiber of flyers, and then mixed and sown by a tractor seeder. In this way, it is possible to place seeds taken from different parts of the field in the neighborhood. Due to cross-pollination, the viability of their offspring is increased. The seeds obtained from these crops are stored in the warehouse for 5 years. A seed nursery is established on their basis [138. P.10-11, 146. P.262-265].

In order to determine the effectiveness of any techniques commented [122. P.31-33] it is necessary to observe, first of all, the condition of comparability of the obtained results. In this case, elite materials of the corresponding nurseries should be compared.

Seed breeding work should be conducted according to a new, more simplified, but quite effective methodology of work with elite material. This can be the American method of breeder Smith, the so-called system of seed reservation, or its modification [101. C.5-7, 118. C.115-116, 128. C.98-99, 139. C.274-276, 146. C.262-265, 183. C.336-339, 187. C.89-91, 210. C.51-52, 241. P.89-90], assuming reproduction of the elite once every 5 years. In this case, it is much easier to preserve germination of elite seeds. In addition, in conditions of rapid change of varieties, procurement of expensive seeds is simply not economically justified.

Based on the concept of degeneration of self-pollinated varieties [121. P.7-12], it is suggested to increase the volume of line sampling in order not to impoverish the heredity of varieties, to conduct intensive scraping and long-term testing in order to exclude the "worst" offspring in the elite. In the process of using even a well-selected variety, the economic and biological traits peculiar to the variety gradually decrease and it deteriorates. This is due to mechanical and biological contamination, splitting and increase of seed-borne diseases.

The task was set to study the degree of variability of some economically valuable traits in cotton varieties under conditions of self-pollination, cross-pollination and open flowering. It was shown that variety heterogeneity in qualitative dominant traits is revealed in the first two years of self-pollination [68. P.339, 146. P.262-265, 147. P.289-293, 255. P.26-30].

Due to mechanical and biological contamination emphasizes [121. P.7-12] there is a "degeneration" of cotton variety. Loss of resistance to Verticillosis, spontaneous mutation, and in the process of seed aging deviant forms appear, and sooner or later the variety is removed from the register, so breeding and seed production of cotton varieties combining a complex of economically valuable traits with high adaptive capacity to specific agro-ecological zones should be considered as a continuous process [273. P.3-7].

Giving special importance to the methods of breeding process[100. P.21-23, 248. P.140], emphasizes more effective methods of creation and identification of new varieties of cotton, as crossing, selection and others.

When growing seeds in primary seed production units, it is necessary to use all the most effective means (indo-selection, spatial isolation, seed disinfection), which can completely localize diseases and prevent their penetration through seeds into production crops of the variety, as the above causes reduce economic and biological properties of the variety, especially in production conditions [26. P. 10-11, 175. P. 103, 205. P. 437-442, 207. P. 17-18].

In our country, there are mainly three sowing methods practiced: often nested, row (pubescent seeds) and precise, with a given number of seeds in the hole (bare seeds and small pubescent seeds).

Charred - delinted seeds have long attracted the attention of researchers due to better swelling, friability and faster germination.

Under favorable conditions, bare seeds germinate faster, develop better and give an increased yield of raw cotton. However, in years with unfavorable weather conditions, i.e. with high humidity and low temperature, they rot faster, sprouts are thinned and weakened [53. P.21-22, 129. P.100-102, 187. P.89-91,,203.P.4-6, 254. P.110-112].

An important part of this work is to increase the field germination of seeds under unfavorable weather conditions, as it is the case in some years in many regions, after the completion of sowing, prolonged precipitation led to reseeding or thinning of crops. While the encapsulated seeds with chitosan preparation were more resistant to unfavorable weather conditions than non-encapsulated seeds. The preparation UZHITAN has a pronounced stimulating activity, which has a positive effect on the growth of cotton, which is expressed by accelerating the germination of seeds, increasing the height of plants, a significant increase in yield, contributing to the early opening of bolls and obtaining quality raw cotton. In the end, the use of encapsulated seeds for sowing helps to reduce their consumption and increase the seed multiplication rate [142. P. 10-14, 202. P.16-17, 236. P.274-278].

Studies show that the same variety can be selected for multiplication for different traits, which can in generations greatly change the parameters of the variety, in contrast to the parameters described by the author. Therefore, the author of the variety should be directly involved in the selection of seed material in elite seed farms. Elite material from different farms is very relevant and aimed at selecting the best farms growing material that corresponds to the author's description. Elite material from different farms has significant changes in economic qualities and technological properties of fiber that they can be recognized as a new variety. Therefore, the study of elite material grown in different elite farms is very relevant and is aimed at selecting the best farms growing material that corresponds to the author's description. This will allow to preserve the longevity of the variety and ensure the release of fiber corresponding to the world requirements [182. P.39,183. P.336-339].

The application of methods: genetics, population genetics and breeding, under constant seed production control - open wide prospects for scientific activity on creation of new high-yielding, high-yielding, with high fiber yield and quality, ecologically clean, not requiring chemical treatment, highly resistant to extreme environmental factors, with increased economic efficiency - cotton varieties; and production of genetically homogeneous, on morpho-economic traits of cotton, pure-seeded seeds of new varieties to ensure the development of new varieties of cotton for the production of new varieties.

The provision of seeds should not be through redistribution, but on a market basis.

The market of agricultural seeds is a complex system, where the following should be involved: state structures of republican and regional level, breeders, seed growers, seed processors and their consumers [126. P. 91-93, 136. P. 64-66, 243. P. 4-7].

It is necessary to oust from the seed market those who are not worthy to engage in seed production, and to create optimal legal conditions for seed production and breeding, institutions and enterprises of any form of ownership, guaranteeing the production of high quality seed [136. P.64-66, 139. P.274-276].

Under current conditions, in order for breeding and seed production to become widely available and cost-effective means of restructuring agriculture, a new variety must be a carrier of economic growth. Only the generation of commercial (highly profitable) varieties can lead to the fact that the carrier of economic growth will be the entire structure of cotton production.[139. P.274-276, 140. P.30-32]

Considering the economic perspective in seed production [215. P.229], emphasizes that economic efficiency in seed production can be achieved as a result of consistent implementation and development of a number of important activities to create a chain of "Breeding, seed production and multiplication", their relationships in links and structures in the direction of ensuring the balance of supply and demand. It is not enough to set settlement prices for seeds only on the basis of cost price. The category of the latter does not take into account fund intensity, labor intensity and other factors.

The activities of seed producers should be directed first and foremost at ensuring seed quality assurance.

The Law of the Republic of Uzbekistan "On Seed Production" [2] recognizes certification of seeds and planting material of agricultural crops as one of the most important conditions for ensuring quality assurance of seeds. It is aimed at permanent control over production, harvesting, processing, storage, sale, transportation and use of seeds, as well as at harmonization of the certification process with the rules and requirements of international organizations and similar systems of foreign countries [206. P.18-21].

According to international practice, the certification process should include: application for certification; application review and decision making; control over compliance with standards and other regulatory documents in production, processing, packaging and marketing of seeds; varietal identification; sampling for testing; analysis of received materials and decision making on the possibility of issuing a certificate; inspection control of certified seeds [127. P.93-96,133. P.19-20].

Particular attention is paid to the varietal qualities of the seeds, through the application of field inspection, to be sure that appropriate technical techniques are carried out to guarantee the preservation of the variety [93.P.218, 94.P.470, 125. P.62-65, 137. P.17-18].

Characterizing the theory and practice of seed production, it is emphasized that each variety of cultivated plants is characterized by a certain set of useful properties and traits for humans. Preservation of these useful qualities of a variety in the process of its multiplication and production use is the main content of seed production work [119. C.21-23].

To solve this problem, it is necessary to develop new and further improve existing methods and techniques of seed production. This, in turn, necessitates deeper and deeper study of the nature of those biological phenomena, the skillful use of which predetermines the successful solution of modern seed production problems [216. P.5-7, 241. P.89-90, 268. P.7-11].

Methods of elite-seed work do not always allow to quickly solve the problem of stabilization of new breeding material obtained by a set of traits, especially since the use of directed selection on several traits without taking into account the rest, in the presence of negative correlation, inevitably leads to weakening or deterioration of other important for the variety qualities.

In plants selected for the earliest opening of the first boll with fruiting not lower than average, there was no definite dependence of raw seed cotton yield on the timing of the first flower and first boll opening. At seed cotton yields from 20 to 80 g per bush, the difference in the onset of flowering was within ± 1 day by class, and the difference in the onset of boll opening ± 2 days. Consequently, individual selection of average fruiting plants by the beginning of boll opening, and even more so, selection without taking into account the total fruiting of bushes does not guarantee a full yield of raw seed cotton.

When evaluating plants for individual selection at one time before harvest, the main attention of seed growers of elite farms is often focused on the number of bolls opened by the time of screening. Therefore, with such a principle of selecting plants for selection, their viewing at earlier calendar dates will be associated with an increase in the number of low-weight individual selections [190. P. 36-39, 194. P. 8-9].

Individual selection of only typical plants by modal selection technique allows to increase the reliability of selection of more productive plants, which helps to maintain and improve the productivity of the variety.

Selection of the best families and plants by phenotype according to the method of selection by correlation of traits partially involves in further reproduction of elite heterosis hybrids that are not detected in field observations, which reduces the reliability of selection of elite materials typical of the variety by economically valuable traits [194. P.

8-9, 262. P.52-57].

The most acceptable method of original seed preparation may be the "Method of field cleaning from atypical plants". Three years of research have shown that this method can improve the varietal performance of seeds. This method can serve as a basic method for creation of the final method of production of original seeds.[146. P. 262-265, 192. P. 15-16, 193. P. 25].

The development of the theory and practice of seed production has a long history. And although notable results have been achieved on this path, the list of problems that remain unresolved is still quite large. The theory of seed production now looks like a mechanical summary of various concepts and facts, assumptions and hypotheses, often insufficiently substantiated and contradictory [40. C.34-37, 41. C.32-34, 99. C.26-29, 135. C.11-12 , 204. C.6-9, 210. C.51-52].

§1.3 Analysis of the influence of seed quality on sowing, field, yield and economic-valuable qualities of plants

It is known that a seed is formed from a fertilized testa. The new organism includes hereditary possibilities of two organisms - maternal and paternal, i.e. during sexual process a new organism with enriched heredity is created. Since gametes are not the same, they are carriers of different hereditary possibilities. This creates genetic prerequisites for the emergence of hereditary heterogeneity of seeds. This is the essence of genetic heterogeneity of seeds, which conditions the genetic construction of the embryo. It affects the longevity, duration of maturation, attitude to various environmental factors [29.P.3-8, 35. C. 43-77, 43. C.112-116, 69. C.282-287, 86. C. 3-6, 87. C.587, 161. C. 97, 235. C. 33-35, 279. C. 117-128, 287. C. 6-26, 288. C. 1365-1368, 289. C.120-124]. Each seed, as it is known, has its biological differences individuality. These differences are morphological and physiological. Even within the most aligned variety of self-pollinated crops, each seed is biologically different from others, although, in general, it retains the main features of this variety and its character of metabolism [226. P.

26, 263. P.187].

Studies conducted in different years have established that early maturity is one of the important economic and valuable indicators of cotton. Many domestic and foreign researchers have studied the nature of inheritance and variability of this trait [9. C.5-9,13. C.23, 14. C.34-36, 38. C.43-46, 59. C. 29-30, 69. C. 282-287, 70. C.47, 74. P.79-80, 79.P.22-28, 90. C.54, 110. C.110, 118. C.115-116, 156. C.87-89, 170. C.18., 181., 231. C.120-121, 238. C.24, 240. C.130-132, 244. C.17-20, 245. C.24-31, 246. C.221, 247. C.22, 265. C.3-6, 280. , 293. C.353-364].

Such seed production techniques as intra-sort pollination, additional pollination, etc. can be referred to the use in practice of genetic diversity of seeds. It is established that after 5-6 years of storage germination of tomato seeds from self-pollination is 33%, from intra-sort 86,5% and from inter-sort-87% [42.P. 76-81, 98. C. ,42. C.57-58,112. P.21,141.P.12-13, 223. C.18-19, 299. C. 19].

Genetic diversity of seeds also occurs at different ages of reproductive organs, their different physiological ripeness. Physiological processes in such seeds proceed with unequal intensity, which leads to different accumulation of metabolites and changes the sowing qualities of seeds [114. P.22,123. P.40-43,152. P.101-105,285. P.108-112].

The phenomenon of matrical diversity of seeds of different crops is well known, which is caused by the place of their formation on the mother plant [44.P.19-20, 56. C. 38-39, 63. C.36,75. C. 30-31, 151. C. 17-19, 177. C. 121-123, 178. C. 54-86, 225. C. 24, 230. C. 304-325].

As indicated [154. P. 108-112, 228. P. 30-70] seed growth on the mother plant is characterized by accumulation of dry matter, and due to a decrease in moisture content - by a decrease and then an increase in raw mass. During this period, the plant is affected by many factors related to the life activity of the mother plant and the diversity of seeds formed.

The place of seed formation on a plant determines their

heterogeneity in a number of traits. This is caused both by the biological properties of reproductive organs and the supply of plastic substances to the forming seeds. The yield and quality of seeds are associated, first of all, with the structure of the bush, i.e. its architectonics [32. C. 207-208, 48. C. 82. C. 4-6, 174. C.33-38, 188. C. 6, 229.P.26-34, 282.P.69-75].

The phenomenon of ecological diversity is very important in the production of high quality seeds, as the yield and quality of seeds largely depend on soil and climatic conditions, applied agrotechnics [261. P. 104-106, 269. P. 5-38]. The conditions of seed formation change their chemical composition and physiological properties. Therefore, seeds grown in different soil and climatic conditions, under different cultivation technology, have different physiological, sowing and yielding qualities, i.e. they are heterogeneous, This heterogeneity determines the ecological diversity of seeds and its use in seed production. There are many publications on ecological diversity of vegetable seeds in the literature.

Different agroecological conditions have a significant impact on seed morphogenesis, which should be taken into account when evaluating and selecting the ecological niche of cultivation in order to obtain a stable yield with high sowing qualities. At the same time, some traits are more stable or vary little under the influence of environmental conditions (plant size, elements of productivity and seed quality, duration of ontogenesis periods), while others change more strongly and depend more on the fluctuations of external factors [20. C. 19-21, 22. C. 129-132, 26,27. C. 27-29, 28. C. 4-11, 43. C. 112-116, 46. C. 84-93, 47. C. 15-16, 48. C. 112-115, 54. C. 23, 198. C. 298-300, 242. C. 31].

In some cases, under an unfavorable combination of external factors, the potential capabilities of plants are practically not realized, and productivity is reduced to a minimum. Under such conditions, studies in the soil-climate-plant system require a thorough analysis of the environmental situation. It is also very important to take into account the microclimate conditions [55. C. 19-20, 74. C. 79-80, 76. C. 49, 92. C. 260, 99. C. 26-29, 195. C. 20-23, 221. C.98].

The most important aspects of adaptive seed production and its peculiarities are: taking into account the differentiating capacity of the environment in the seed production zone; taking into account the adaptive properties of the variety at the location of its seed production; optimal combination of the reproduction zone and the zone of seed realization. She also points out that the impact of the environment on the plant can be stabilizing in the phases of vegetative development of the plant up to technical ripeness and destabilizing in the period of seed formation [71. P. 10-11,72. P. 11-13].

Intensification of agricultural production naturally leads to concentration of seed crops and concentration of seed production in certain areas of the country. It is very important to find out what seed yields can be obtained in different zones and how ecological conditions affect progeny productivity when seeds are used in different soil and climatic zones. Information on the productivity of seed crops in different ecological conditions and on seed yield qualities is contradictory.[222.].

The Government of the Republic of Uzbekistan in connection with significant shortcomings in the work on cotton seed production on November 28, 1998 № 491 adopted a Decree, which states, in particular, that the system of seed production is not properly worked out, and the transition to market relations has not been implemented. It was also noted that the requirements of the Laws of the Republic of Uzbekistan "On Seed Production"[2] and "On Breeding Achievements" [3] are not duly observed, and the system of seed production is not established. [3], cooperation between production and science in the field of seed production is not established [4].

Taking into account the requirements of the Government, a tender for the right to produce raw seed cotton and registration of physical and legal entities producing elite and subsequent reproductions has been organized [145.P. 22, 197. P. 83].

Some authors [21. P. 251, 28. P. 4-11, 43. P. 112-116, 149. P.33-36, 267. P.14-17] note that single or double reproduction of seeds in new conditions does not cause noticeable practically important changes in varietal traits. In the opinion of others [217. P.231, 267. P.14-17, 276. P.11-15], cultivation of seeds in environmental conditions not typical for a given crop inevitably leads to profound biological changes in the plant and to the loss of valuable traits in the offspring. Moreover, this influence can be both positive and negative. A number of researchers point out that local reproductions of seeds are always better than imported ones [23. P. 194-195, 54. P. 23, 148. P. 115, 207. P. 17-18, 237. P. 6-7].

At the same time, there is evidence of insignificant changes that occur during single and multiple reproduction of seeds under new conditions, especially if soil and climatic zones are favorable for seed production of this crop. In some cases, such changes lead to increased seed productivity and improved quality of marketable vegetables obtained from these seeds. Transfer of seeds of heat-loving crops from the southern zone to the north contributes to higher yields of marketable vegetables [31. C.57-65, 36. C.35-37 , 112. C.21 , 130. C.109-110, 132. C.103-104, 150. C.29-33, 161. C.97 , 163. C.31 , 207. C.17-18 ,218, 220. C.106-116, 224. C.177-179.,239. C.19-22., 246. C.221 ,267. C.14-17].

The influence of the place of seed reproduction on the productivity of offspring and the need to organize zonal seed production of vegetable crops is indicated by many other researchers [35.P.43-47 , 36.P. 35-37, 171.P.15-26, 172.P. 5-7, 227.P. 78-79, 284.P.12, 107 , 286.P.29-32 , 291. C.249-256].

There is no unanimous opinion on the criterion for indirect assessment of seed yield properties. Some argue that the best yielding properties have seeds with higher mass and reduced weight [57. P.20-21 , 173.P.16-18 , 186. C.20-23 , 208. C.12-15 , 219. C. 160, 220. C.106-116 , 233. C. 28], others, on the contrary, - with less [26. P. 10-11, 216. P. 5-7, 253. P. 21-25, 277. P. 447, 278. P. 117-123], others -

with better sowing properties [180. P. 29-31, 187. P. 89-91, 188. P. 6 , 189. P. 103-112], others - with a certain chemical composition [30. P. 106-108]. Such differences in views, in our opinion, are due to the fact that the authors of the publications based their conclusions on the results of individual experiments, working in an ecological zone where the studied factor could have a significant effect on the productivity of offspring.

Under the influence of environmental conditions, not only physical but also sowing properties of seeds change to a great extent. The germination energy and germination capacity of seeds depend to a greater extent on the correct process of their receipt and further storage [46.P.84-93 , 166.P.69-76 , 196.P.23-25 , 209.P.37-38, 217. C.231., 232. C.167, 281. C.12].

Any seed lot is formed from a huge number of different quality (heterogeneous) seeds. Seeds included in a seed lot have different moisture content, chemical composition, different responsiveness to environmental conditions, as well as unequal productivity. Different quality or heterogeneity of seeds in a sowing batch is an objective regularity. This phenomenon is typically biological, depending on many factors: heterogeneity of morphogenesis stages, unequal sex elements involved in fertilization, anatomical structure of the conducting system, differences in the activity of the assimilation apparatus, mineral nutrition and water supply [63.P. 36,144.P.477].

Because seeds are biological products, their behavior cannot be predicted with the same certainty as in the testing of non-biological products. The exchange of seeds between countries and areas requires that the conditions of testing in one laboratory coincide with the conditions of testing in another laboratory [259. P. 19-20].

The above literature review indicates that seeds formed on different parts of plants, as well as grown under different soil and climatic conditions and under different agrotechniques, are not homogeneous in chemical composition, physiological and biochemical properties, physical, sowing and yield qualities.

The present work is aimed at solving actual problems of cotton seed production and improving methods of seed quality assessment.

CHAPTER II. LOCATION, CONDITIONS, MATERIALS AND METHODOLOGY RESEARCH ACTIVITIES

§2.1 Location and conditions of the studies

Studies on creation of computer program were conducted at the Republican station of primary seed production and seed science of agricultural crops (now Research Institute of breeding and seed production and agro-technology of cotton cultivation) by laying mock experiments on varieties sown in elite farms in order to compare the existing and computer evaluation of seed material, including time, cost, cost-effectiveness, as well as purity and homogeneity of elite material. Another part of the research was conducted in elite farms of Tashkent, Surkhandarya, Samarkand, Fergana, Andijan and Jizzak regions, the basis of the research were taken the main indicators of fiber quality, criteria for rejection and selection of elite material for further multiplication, taking into account the correlations of economic-valuable indicators.

The Institute is located on the territory of Kibray district of Tashkent oblast, at the 42nd degree of northern latitude, at an altitude of 481/m above sea level. Soil - typical, cultivated serozem of long-standing irrigation, from the main canal Boz-suv. Soil data and characterization are given in Tables 1 and 2.

Table 1

Soil mechanical composition

Depth cm	Fractional volume, mm							
	1-0,25	0,25-0,1	0,1-0,5	0,05-0,01	0,01-0,005	0,005-0,001	0,001	0,01
0-30	7,29	2,43	4,64	34,5	15,16	18,82	17,16	51,14
30-50	7,34	1,63	11,27	36,72	11,60	15,6	15,84	43,05

Table 2

Soil fertility

Depth	Total number			Movable form, mm		
cm	Humus	N	P	N-NO3	P2O3	K2 O
0-30	0,918	0,088	0,118	24,2	31,6	140,0
30-50	0,856	0,079	0,100	20,4	27,6	120,0

Mechanical composition of the soil is heavy loam, and groundwater is located at a depth of 12-15 meters. In arable layer 0-30 cm. humus content is 0,918%, in subsoil layer 30-50 cm. humus contains 0,856%, i.e. the difference is not big. The content of nitrogen and phosphate in the arable layer 0-30 cm is 0.88-0.79%, and subsoil 30-50 cm-0.118-00.70%.

The amount of nitrate nitrogen, suspended phosphorus and potassium respectively is 26.4-21.4; 31.6-27.6; 140-120 mg/kg, i.e. the experimental field belongs to the number of medium fertile soils.

At an average temperature below 10°, medium-fiber cotton suspends its development. Therefore, temperatures above 10° are considered effective. The sum of effective temperatures at the lower limit of 10° from sprouting to budding is 400-420°, from budding to flowering 500°, from flowering to opening of the first bolls-875-885°.

The sum of effective temperatures from sowing to opening of the first boll in medium-fiber varieties is 1875-1885°, for thin-fiber late maturing varieties-1955-1965°, for early maturing-1720-1730°.

The source of water supply is permanent irrigation ditches of on-farm use originating from the Boz-Su canal. Meteorological conditions of the area are typical for sharply continental climate and are characterized by cloudless weather, providing high intensity of sunlight and characterized by insignificant amount of precipitation.

According to long-term data, the average annual precipitation is about 360 mm, mainly in the fall-winter-spring period. The period of mass fruiting of cotton is marked by minimum precipitation, low atmospheric humidity and the highest air temperature.

According to long-term data, the last spring frost usually occurs in late March and the first autumn frost in the second decade of October. Late spring frosts rarely occur in April and very rarely in early May. The earliest fall frosts are observed in early September and the latest in mid to late November. The vegetation periods during the field experiments were generally stable and favorable for normal growth and development of cotton.

Temperature and precipitation conditions for a number of years are shown in Table 3.

Agrotechnical measures on the experiments were carried out according to the terms and methods of agrotechnical methods accepted at the station. Annually in November-December, winter plowing was carried out with application of 50% of the annual rate of phosphorus fertilizers. In early spring harrowing was carried out, then chiseling, before sowing harrowing with malovanie. In the second half of April sowing was carried out and, if necessary, feeding irrigation. As weeds appeared, two to three times hoeing was carried out. During vegetation 5-6 cultivations, two with fertilizers (nitrogen, phosphorus, potassium), 3-4 irrigations of 800-900 m^3 of irrigation water were carried out. Harvesting was carried out in September, October and finished not later than November 10.

Table 3

Average daily temperature and precipitation for 2010-2014

(according to data of Bozsui meteorological station)

Months	Average daily air temperature (C)^0					Average daily precipitation (mm)				
	2010	2011	2012	2013	2014	2010	2011	2012	2013	2014
April	13,5	15,1	15,9	16,8	14,2	32	11,3	39,2	33,5	38
May	18,7	22,3	20,0	23,2	19,0	22	6	32,6	21,0	25,3
June	24,5	27,9	27,2	26,5	24,8	9	0,4	3,8	9,2	25
July	27,8	27,2	28,7	27,4	26,8	-	7	0,3	2,2	-

August	26,2	27,1	25,5	27,1	27	-	-	2,7	-	0,4
September	21,2	21,8	22,5	20,3	15,1	-	1	-	0,2	-
October	16,0	13,0	15,0	17,0	17,4	5,5	6	4,2	38	-
November	8,2	11,4	8,2	9,4	9,9	33	22	50	73,1	
Medium	19,6	20,7	19,8	20,9	19,3	12,7	6,7	16,6	22,1	12,0

§2.2.Research methodology

Domestic methods were applied in the research on determination of sowing qualities of cotton seeds.

For germination determination, seeds were germinated in sand and filter paper rolls. The sand for analysis was sieved sequentially through sieves with aperture diameters of 1.0 and 0.5 mm. The sand remaining on the second sieve was washed until clear water drained out. The washed sand was calcined until a piece of paper placed in the sand was charred, after which the sand was again sieved through a sieve with 0.5 mm holes.

The germinated seeds were counted on 4-12 days. At each count, the number of normally and abnormally germinated seeds, as well as diseased seedlings were counted separately. Seed germination on corrugated paper was carried out according to the following method: two layers of paper 100-105 cm long and 12 cm wide were corrugated so that there were 24-25 folds with the height of teeth 20-22 mm each. The paper corrugated in this way is moistened, placed in a planting bed and 4-5 seeds are placed in each fold.

Varietal purity was determined in accordance with the instruction on cotton approbation, approved by the MAWR of RUz in 2002, once in August, and morphological features were taken as a basis:

(a) The size, pubescence and shape of the leaf;

b) pubescence of the main stem;

c) branching type and bush shape;

d) size and shape of bolls.

The determination of varietal qualities during the inspection check started with familiarization with the crop, and first of all attention was paid to whether it corresponded to the official description.

In order to master different methods of production of super elite, elite and first reproduction seeds, research work was carried out in field and laboratory conditions in elite seed farms: Chinaz and Yukoro-Chirchik variety C-6524, Tashkent region, as well as at the Republican station of primary seed production and seed science of agricultural crops. Seed production according to the existing methodology was taken as a control.

In nurseries of 1 and 2 years, seed production work was carried out according to the current methodology. On the remaining, after field rejections, families, samples, individual selections and seed collections were collected, on which laboratory evaluation and purification on DL-10 gins was carried out, a sample was taken from fiber for analysis of technological properties, and seeds were used for planting of super-elite crops.

From the seed multiplication nursery, seed raw cotton was selected from the remaining families and plants after field rejections and delivered to the cotton ginning plant for preparation of elite seeds.

Raw seed cotton was also collected from the super-elite crops, run through DL-10 laboratory gins, samples were taken from the fiber for HVI analysis, and the seed was used for the elite crops.

In the field experiments, observations of plant growth and development were carried out, varietal purity, disease and pest infestation, home-grown and total yields were determined. In laboratory analyses were determined indicators provided by the current instruction for the production of seeds of elite and first reproduction of cotton varieties (1981) and the instruction for the preliminary propagation of new varieties of cotton. (1986 г).

In order to substantiate the indicators of the conducted research, the methodological basis for assessing the materiality of the difference between the compared variants, confidence interval and correlation relationship according to B.A. Dospekhov (1985) [83] was used, using the formulas:

1.Estimation of the difference of mean of independent samples with the same number of observations $n_1 = n_2$

$$S_d = \sqrt{S_{\ddot{x}1}^{\ 2} + S_{\ddot{x}2}^{\ 2}} \ , (1)$$

where S_d - errors of compared arithmetic averages $\ddot{x}_1 \ u \ \ddot{x}_2$

$$S_{\ddot{x}} = \frac{\sum (x - \ddot{x})^2}{n-1} \qquad (2)$$

where $\ddot{x} = \frac{\sum x}{n}$ is the arithmetic mean; n-number of observations

2.Materiality of difference criterion

$$t = \frac{\ddot{x}_1 - \ddot{x}_2}{\sqrt{S^2_{\ddot{x}1} + S^2_{\ \ddot{x}2}}} = \frac{d}{S_d} \ (3)$$

3.Estimation of materiality of the mean difference (conjugate samples) by difference method

$$S_{\ddot{d}} = \sqrt{\frac{\sum (d - \ddot{d})^2}{n(n-1)}} \quad или \quad S_{\ddot{d}} = \sqrt{\frac{\sum d^2 - (\sum d)^2 \cdot n}{n(n-1)}} \ (4)$$

where n is the number of conjugate pairs;

d-difference between conjugate pairs of observations.

Highest significant difference for the significance level of 0.95

$$HCP_{05} = t_{05} \cdot S_d \ (5)$$

where t is Student's criterion

Confidence interval

$$\ddot{x} \pm t_{05} \cdot S_{\ddot{x}} \ (6)$$

4.Estimation of correlation closeness

$$\tau = \frac{\sum (x - \ddot{x})\,(y - \ddot{y})}{\sqrt{\sum (x - \ddot{x})^2 \cdot \sum (y - \ddot{y})^2}} \ (7)$$

where x is an independent variable;

y-dependent variable;

$\ddot{x}, \ddot{y}$ -arithmetic averages for the series x and y

Correlation error

$$S_{\tau} = \sqrt{\frac{1-\tau^2}{n-2}} \quad (8)$$

where τ is the correlation coefficient;
n-number of samples;
S_{τ} -correlation coefficient error.

5. Regression equation

$$y = \ddot{y} - B_{yx}\,(x - \ddot{x}) \quad (9)$$

where B_{yx} - regression coefficient

$$B_{yx} = \frac{\sum (x-\ddot{x})\,(y-\ddot{y})}{\sum (x-\ddot{x})^2} \quad (10)$$

§2.3.Background material used in the research

Seeds of cotton varieties: C-6524, Sultan, An-Bayaut-2, Bukhara-6, Namangan77, which are included in the State Register of agricultural crops recommended for sowing on the territory of the Republic of Uzbekistan, were used as a source material for research.

The following chapters discuss the impact on improving the reproduction of super-elite material by developing different seed production methods.

CHAPTER III. COMPARATIVE STUDY OF APPLIED METHODS OF ELITE SEED PRODUCTION FOR THE PURPOSE OF THEIR USE IN IMPROVING THE COTTON SEED PRODUCTION SYSTEM IN THE REPUBLIC

§3.1 Comparison of applied methods of primary cotton seed production in Uzbekistan and abroad

The beginning of the history of domestic breeding and seed production dates back to the second half of the 19th century, when the first attempt was made to import American cotton seeds. Until that time, Uzbekistan had been cultivating local gooses belonging to G.herbaceum and G.arboreum L. They were low-yielding, had small bolls with low yield-25-30%, coarse and short (18-25 mm) fiber (Efimenko, 1976).

Imported in 1870, the American cotton plant C-Island was not adapted to the natural conditions of Uzbekistan due to its late maturity. Therefore, in 1878, seeds of faster maturing varieties were imported: King, Roussel, Cleveland and others. Since then, American varieties have gradually started to replace local guzy (Encyclopedia of Cotton Growing, 1985). The lack of established breeding and seed production work led to the fact that the raw cotton yields of different varieties were taken without any accounting, which led to the mixing of raw cotton and the formation of factory mixtures.

The plant mixture in the process of cultivation in different soil-climatic zones of cotton sowing acquired special economically valuable qualities and by the name of administrative areas of cultivation became Bukhara, Khorezm, Chimbay, Kokand and others.

1909-1910 should be considered as the beginning of scientifically based research in experimental institutions. At that time, breeding departments were organized at Golodnostepskaya and Andijan stations, where varieties Navrotsky, Triumph of Navrotsky, Dehkan, Akdzhura, 169 and 182 were developed. However, due to the lack of seed breeding work, multiplication of varieties and their introduction into production was extremely slow (Encyclopedia of Cotton Production, 1985).

By 1928, selected varieties of Navrotsky, Triumph of Navrotsky, 169 and 182 on the main areas had displaced the inbred seed stock of factory mixtures and guzes. However, multisorting within farms as a result of poor seed production resulted in rapid variety contamination (Alexandrov, M, 1962).

Analysis of the reports of the Central Control and Seed Cotton Station (CCSCS) for 1934-1948 showed that in 1884 only 300 dessiatinas were occupied by varieties of G hirsutum L species near Tashkent. Good results were obtained for these varieties.

Natural selection led to the emergence of varieties characterized by rapid maturity and higher economic and technological properties (Alexandrov, 1949). This was the beginning of scientifically-based work on cotton seed production.

In connection with the replacement of cotton variety mixtures with purely varietal ones, in the 30s of the last century, elite-seed breeding work was organized, based on an individual-mass approach to the evaluation and selection of elite material. Any refined variety in the most distant reproductions (reseeding of elite) persistently retains its economic and biological traits that determine yield. Undesirable changes in heredity for these and other traits are not group, but individual, they occur rarely, only in individual plants and therefore can not quickly, without a very long action of natural selection to affect the deterioration of the variety. The regionalized variety was transferred for multiplication to the elite farm, where its multiplication and varietal renewal in seed farms was carried out without the participation of the authors of the variety.

In varietal renewal worldwide, seed production and improvement of commercial qualities of cotton varieties are handled by special seed companies, while government research institutions develop seed production methods. Although there are no uniform methods in practice, each breeder-originator or relevant research institution produces annually the required amount of elite seeds by the method they consider acceptable. (Table 3).

Cotton seed production in the USA is specialized to the seeds, the following requirements are presented: germination 80-85% and higher, purity of crops by morphological features not less than 80%, relative uniformity of cotton fiber by technological properties inherent to the variety. The company covers all costs if the varietal purity and field germination of the sold seed is lower than the norm. High purity is achieved by the fact that, firstly, the plant purifies only one variety, secondly, in the field, spatial isolation of crops is observed, several varieties are never sown in one farm, thirdly, ruthless culling of atypical plants is carried out, and fourthly, the author's team is fully responsible for this work. Under his leadership, the work on the production of high-quality seeds is organized. Breeding and seed production work is built as one inseparable process. The authors of the varieties are in charge of seed production.

Table 3

Comparative table on elite seed production in Uzbekistan and USA

1.	Method applied in Uzbekistan without application of WSS	Nursery I year		Nursery Year II	Seed propagation nursery		Reproduction in the crop	Some methods used in the United States		
								Method name	1 year	Year 2
		1500 i.o. area 1.08 ha		400 families area 2.88	250 families area 30-36 ha		R-1-R-3	Mass selection of a large number of typical plants	Selection of 40.0 thousand I.O. and bracing	Vysev I.O. as elite
	Method with the application of VSS	VSS Kennel		Nursery 2 years	seed propagation			Field clearing of atypical plants	Stock 10 thousand I.O.	Kennel of super elite
		paternal part	mother part		nursery study 1.80 ha	seed nursery 35-40 ha	R-1-R-3	Modal pattern system	6.0 thousand E.O.	The original nursery
		400 i.o. area 0.29	600 i.o. area 0.43	Nursery area 2.9 ha				Progeny row testing	Selection and rejection E.O.	Nursery of 1 year
								Repeated trials of progeny lines and lines	1000 E.O. Nursery 1 year	Nursery 2 years

								The Pedigra system	1000 I.O. field and lab braking	Field and laboratory culling, selection of the best families
								Seed reserve or storage system	Harvesting of I.O. - obtaining 1000 kg of seeds	1000kg of seeds- original nursery

Fields or plots for stock and registered seed production shall be isolated 1320 feet (one foot-30.5 cm) from another variety of the same type or 2640 feet+20 additional protective rows from other varieties of different types.

Fields where certified seed is produced must be isolated 660 feet+20 additional protections from other varieties of different types or 20 feet from other varieties of the same type.

For separate production of stock, registered and certified seed, fields and their plots are not isolated according to requirements. In these cases, a defined boundary such as a fence, ditch, road, dam or protective strip is required. Any cotton variety, in a partition of an isolated distance, must be objectively identified.

Fields for certification should be inspected at least once before collection, after opening several boxes. Each field

must be cleared of other plants, varieties. Maximum mixture of other

varieties or atypical varieties are allowed: stock and registered - one plant per 45,000 and certified - one plant per 9,000.

For stock and registered seed production, land where cotton was grown the previous year is not suitable if even cotton of the same variety is grown for certification and meets the inspection requirements for varietal purity. No different types of varieties shall be sown in fields for the production of certified grade seed. Fields shall be free of all prohibited noxious weeds that are difficult to identify. Fields in unsatisfactory condition must be abandoned.

Elite seeds are multiplied directly in the farm of the seed company or by contract in individual cotton farms. Seeds of 1st reproduction, then 2nd and 3rd are multiplied by the same system. For mass sowing, seeds of 3 reproductions are mainly used. A significant part of seeds of the 4th and subsequent reproductions is used for technical processing. However, some farmers, for economic reasons, use their own seeds of 4 and subsequent reproductions in their crops, but this is an exception. Farmers buy more than 85% of sown cotton seeds from specialized seed companies. In the elite plots, diseased and underdeveloped plants atypical for the variety are removed and field trials are conducted. In the fields of 1 and 2 reproductions, field testing is also carried out, and the crop is harvested by machines.

Elite seeds, as a rule, have 100% varietal purity: 1st reproduction-99%, 2nd reproduction not less than 95%. Seed germination of all these reproductions is not less than 80%.

Up to 85% of the total U.S. seed crop is harvested for seed and the remainder is processed for technical purposes. Seed moisture content does not exceed 12%, but is predominantly 8%. The varieties "Stonville", "Pedigreed Seal" and "Delta and Pine" of the companies occupy large areas in the cotton clade of the USA, Spain, Greece, Turkey, Latin America and others. Such authority is gained only by the fact that the varieties of these firms are of high purity, the seeds are of high germination and purity, and the fiber is in demand in textile enterprises.

Here they duplicate seed crops, renting land to locate in different states. For example, the Stoneville firm sows the same seed in Mississippi and Louisiana for conservation purposes.

Israel has a special seed law. A certificate is issued for sowing seeds of a variety. The main indicator of a variety is yield. A distinction is made between breeding, stock, registered seeds, which roughly correspond to such categories as elite, first, second and third reproduction seeds.

The next section analyzes the development of cotton seed production in Uzbekistan.

§3.2 Development of cotton seed production in Uzbekistan

In our republic, the main tasks in the cultivation of super elite, elite seeds are the following:

a) maintenance of all valuable biological features and economic qualities of the variety;

b) further improvement of these valuable qualities in the seed production process;

c) recovery of seeds from diseases and pests;

d) maintaining high purity;
e) rapid multiplication of seeds for variety change and variety renewal.

This is done by varietal renewal, i.e. by replacing less valuable seeds with high-grade seeds of the same variety of the best reproductions. Elite seeds after sowing give seeds of the first reproduction. Sowing seeds of the first reproduction yields seeds of the second reproduction, etc. Cotton seed production in the 30-40s of the last century was not rid of a number of significant shortcomings, and there was no experimental data to improve the methodology. But gradually the results of work of elite farms were generalized and research data were accumulated in scientific institutions. Studying the wide experience of elite farms made it possible to identify shortcomings in the methodology of varietal seed production and improve it. Thus, in 1940 the following activities were carried out.

Elite work was overloaded by hand sowing large quantities of seeds with up to 10-15 thousand or more individually selected plants left for one elite farm for each variety.

The progeny of seeds collected from individually selected plants in elite crops were checked by production groups, i.e. by second progeny, by comparing the average indicators of economic qualities for the group of families. It was difficult to identify non-typical plants and families in such an averaged evaluation. The total collections from the non-rejected seed multiplication families were pooled together to form the first reproduction of the variety for the future.

From 1951 to 1967, elite work in Uzbekistan was based not only on individual plants, but also on whole families with positive traits, which limited the use of potentialities inherent in the variety. To improve the quality of work, the number of selected plants and families was reduced by 3-4 times.

Individual selections were checked and evaluated by progeny on the basis of families, i.e. the first progeny. This made it possible to reject all the worst families directly from the progeny of each individual selection, leaving the progeny of the best families for further breeding. To prevent a decrease in the number of progenitors and, consequently, degeneration of the variety, 20-25% of individually selected plants from the best, most productive, high-grade cotton crops of the second and subsequent reproductions were added annually.

The changes introduced in the methodology provided a significant reduction in the cost of production of elite seeds and relief from cumbersome technical work. In parallel with the changes in the methodology of selection and evaluation of cotton seed material for economic qualities, the evaluation of each family for technological properties of fiber was strengthened. Laboratory analysis of test samples was introduced to determine the weight of one boll (g), fiber length (mm), lucidity (quantitative and weight method) and fiber yield (%). On the basis of field evaluation data and test sample analyses, we began to reject the families whose basic parameters were worse than the average of all families. The best plants were selected from the best families that were not rejected, and their seeds were harvested separately for further multiplication.

Since 1944, the seed nursery introduced the sowing of standards every 10 families (each family was sown on a separate row). The elite seeds of the same variety released by the elite farm in the previous year were used as standards.

For sowing of seed propagation, seeds were selected only from more yielding and better in terms of the complex of indicators of families in comparison with the standards, and seeds for seed nursery were collected only from individually selected bushes in the best families.

When selecting families and individual plants, in addition to field and laboratory evaluation data of the current year, evaluations of previous years were taken into account.

The elite farms were equipped with the necessary technological equipment, and the heads of elite farms and laboratory technicians in each elite farm were trained in the methodology of analyses.

First reproduction seed production is a responsible link in elite breeding. In addition to the usual approbation, seed production measures in the first reproduction crops included a thorough field inspection of all plants during flowering and culling of all bushes not typical for the variety. At the same time, special attention was paid to the placement of first reproduction crops on fertile plots, instead of concentrating these crops on fields adjacent to elite crops, regardless of their fertility, as it was before.

In the Instructions for Cotton Seed Production issued in 1940-1945,

The main objective was to maximize production of the highest quality, for which elite seeds were grown under conditions of high agro-technique, directed individual selection with verification by progeny. In each elite farm, the following was established: a) seed nursery on the area of 2 ha, where at least 1500 individual selections from 500-700 progenitors and 500 selections from high-yielding crops were sown; b) seed multiplication, where the harvest of seeds of non-rejected families in the seed nursery was sown on the area of 15-20 ha. The total yields from the non-rejected seed multiplication families were pooled together to form the first reproduction of the variety for the future.

From 1951 to 1967, elite work in Uzbekistan was based on the use of intra-varietal crosses. In 1950, 1956, 1963, the Instructions on seed production were based on three principles: cultivation on high agrotechnical background, realization of intra-varietal crosses. In 1960, in Uzbekistan, all sown areas of cotton were sown with seeds that had undergone intra-varietal crossing.

In the experience of L.F. Koloyarova in 1953-1959, observation of the efficiency of intra-varietal crosses was carried out up to the seventh generation, in five generations a yield increase from 6.5 to 1.4% was obtained in the following generations.

Table 4

Observation of the efficiency of intra-varietal crosses

Years	Generation	Reproduction	Deviation,%

1953	F1	-	+6,5
1954	F2	Elite	+5,5
1955	F3	R1	+7,9
1956	F4	R2	+6,8
1957	F5	R3	+1,4
1958	F6	R4	-3,5
1959	F7	R5	-5

Productivity and vitality of varieties, terms of their effective use in production depend on many factors that can be used by breeders and seed breeders.

The method of intra-varietal crossing was a reserve for preserving the initially high heterosis and increasing the vitality of varieties that was not used in cotton seed production. The application of this method in cotton seed production in the first years was complicated and hampered by the recommended extremely labor-intensive work associated with flower castration. It was also unclear how intra-sort crossing affected the technological properties of fiber over a long period of time.

Since 1951, the method of producing seeds of elite cotton varieties with intra-sort crossing without flower castration has been used in all elite farms. However, the possibility of effective use of intra-sort crossing in cotton seed production was questioned by many researchers on the grounds that in experiments they noted an increase in vitality and yield of seed material only up to the second and third reproductions.

Some specialists spoke about the equivalence of low reproduction and elite seeds in many cases. The scheme of varietal renewal according to their proposals was lengthened by 3-4 years. Such prolongation of varietal renewal was unreasonable, it led to the fact that already in 1957. 65% of the cotton crop area in the country was sown with seeds of low reproductions (fifth-eighth), and in large quantities from low-yielding crops. At that time, the best seeds from high-yielding, differently maturing crops were processed at oil mills for technical purposes. At the same time, due to the lengthening of variety renewal, the efficiency of intrasort crossing and improvement of cotton varieties in terms of economic qualities and technological properties of fiber in the process of elite work decreased. Therefore, the joint scientific session on cotton growing, held in 1957 in Tashkent, decided to restore the previous order in cotton seed production and varietal renewal according to the five-year scheme.

In 1966, seed breeders M.V.Lyatsky, A.S.Davshan, P.K.Galitsinsky, A.N.Shafrin and others proposed to remove intra-variety crosses in elite-seed work in Uzbekistan, motivated by the following:

1) the applied method of cotton elite seed production, combining individual directed selection with intra-varietal crossing and elite cultivation on the background of high agrotechnics, was recommended without preliminary testing under production conditions;

2) use of heterogeneous pollen of paternal forms led to heterogeneity of progeny, deterioration of seed material, decrease of homogeneity and increase of deviant forms;

3)labor costs for production of elite seeds increased. Intra-varietal crossing cannot contribute to high yields in several generations.

In 1965, a discussion on the issue of intra-varietal crosses was organized in the journal "Cotton Growing". Both supporters and opponents of intra-variety crosses in cotton seed production were revealed. As a result of the discussion, the Instruction on cotton seed production, published in 1967, envisaged reproduction of elite seeds in two ways: with and without the use of intra-varietal crosses. At the same time, it was assumed that when a variety was zoned, the Ministry of Agriculture, in agreement with the breeding institution or at the suggestion of the variety's author, would establish one or another method of elite seed production.

After 15 years of work in cotton seed breeding with the use of intra-sort crosses, the heredity of cotton varieties of districtized varieties has been greatly shaken, which was repeatedly stated by the workers of elite seed farms at the republican meetings.

For about six years, as paternal forms from which pollen was collected, plants obtained from sowing seeds selected not only from individual plants in other elite farms (irrespective of ecological zones) but also harvested from household crops (irrespective of reproduction) were used, i.e. untested material was used.

Subsequent methodological changes introduced in the instructions of 1956 and 1963 did not improve elite work, as selection of paternal part recommended to be based on the data of elite trials, as well as applications of elite farms. This could not improve the situation, as the experience of "elite-testing", in our opinion, was laid methodologically wrong, without equalized sowing, and applications of elite farms were not and could not be purposeful, as the selection of elites varied depending on the yield indicators of the year and the peculiarities of the season.

This situation was aggravated by the fact that it is impossible to identify atypical forms by the beginning of flowering, as the plants are not fully formed by this time. As a rule, rejection of atypical plants and families is carried out at the second field inspection, i.e. after crossing. It turned out that the pollen collected was not biologically homogeneous and, as a consequence, variegation in the progeny was observed.

In addition, until 1963, the instruction allowed the chasing of plants on the mother part of the nursery of intra-varietal crosses, under the influence of which the shape of the bush sharply changed and the selection of typical plants became difficult. All these facts indicate that intra-sort crosses in cotton seed production were introduced without a sufficiently proven methodology. Therefore, it was proposed to cancel intra-sort crosses in cotton seed production until cotton research organizations prove their efficiency and develop a scientifically grounded methodology of this technique under production conditions.

The basic normative document that currently defines elite work is the "Instruction on production of seeds of elite and first reproduction" of 1981 issue, which legitimized two methods:

1. Individual selection method with progeny testing without combining with intra-varietal crossing.

2. Individual selection method with a combination of intra-varietal crosses.

A careful analysis of the provisions of this instruction shows that it does not provide a clear idea and answer on the issue of guaranteed seed selection.

Until recently, the production of raw cotton seed was carried out by farms that were recommended by regional and district agricultural authorities, without taking into account their material and technical base, staff of seed growers, soil and climatic conditions.

The Government of the Republic of Uzbekistan in connection with significant shortcomings in the work on cotton seed production on November 28, 1998 № 491 adopted a Decree, which particularly stated that the system of seed production is not properly developed, and the transition to market relations has not been implemented. It was also noted that the requirements of the Laws of the Republic of Uzbekistan "On Seed Production" and "On Breeding Achievements" are not properly observed, and cooperation between production and science in the field of seed production is not established.

The developed measures on fulfillment of tasks set by the resolution allowed to reduce the volume of seed preparation for sowing from 181.0 thousand tons in 1998 to 78 thousand tons in 2014, and seed consumption per hectare from 123.7 kg. To 53.5 kg. Thus, the volume of seed preparation for 17 years decreased by 2.3 times, seed consumption per 1 ha by 2.4 times.

Iksanov M, Egamberdiev A.E., Ibragimov P.Sh. (2001) suggest that seed production work should be carried out according to a new, more simplified, but quite effective methodology of work with elite material. This can be the American method of breeder Smith, the so-called system of seed reservation, or its modification.

Sh.Kozubaev (2008) believes that for our conditions, when

technical equipment of elite laboratories is not high enough and in the presence of a shortage of qualified specialists, its modification, assuming reproduction of elite seeds once every 5 years, is more acceptable to begin with. In this case, it is much easier to preserve germination of elite seeds. In addition, in conditions of rapid change of varieties, procurement of expensive seeds is simply not economically justified. Having studied the method proposed by Y.Meredov, which is based on individual selection once in several years, he proposes to modify it and to test it in elite-seed farms of the republic. The method was based on individual selection once in 5 years.

For decades, the question of methods for producing elite cotton seeds has been debated. This is explained by two reasons. The first is that, unlike other crops, the original breeder of a cotton variety does not participate in elite production after its zoning. It is transferred into the hands of seed breeders of elite farms, while for all crops both in our country and abroad (including cotton), every year the variety originator himself prepares elite seeds according to the method he considers most expedient for variety preservation or its improvement.

Then he transfers them to special farms for multiplication. The situation with cotton seed production is quite different. After zoning, the originator of the variety and the breeding institution are released from the annual production of elite, which is transferred to elite seed farms. Specialists of the elite farm have to do a lot of work - select typical bushes, harvest them separately, carry out laboratory work to assess the economic and valuable qualities of raw cotton and technological properties of fiber, cull atypical plants, in general, do everything that is necessary to obtain elite seeds of a variety.

But seed breeders have not been involved in the process of many years of breeding work and do not have an accurate and clear idea of the characteristics of the variety. In addition, new varieties are usually of hybrid origin, so the material given to the seed breeder is usually heterogeneous even in terms of external characteristics. In the process of work it is broken down into groups and it is necessary to decide which of them to take as a basis for a new variety. This work becomes more complicated when several elite farms are engaged in seed production of one variety. In these farms, individual selections, cleaning, and rejections are made by seed growers with different qualifications and different ideas about the best type of variety. Therefore, the same variety may have different branching type, bush habitus, boll size, etc.

Seed production and improvement of commercial qualities of new cotton varieties abroad is carried out by special seed companies. In practice, there are no uniform methods of variety seed production. In particular, in the USA each breeder-originator of a variety or corresponding research institution produces annually the necessary amount of elite seeds by the method, which is considered acceptable. On this basis, the task was set to conduct research on improving the methodology of production of original cotton seeds in the Republic of Uzbekistan.

§3.3 Improvement of methods for multiplication of original seeds of new and registered cotton varieties

In 1986, for the first time, the "Instruction on preliminary seed multiplication of new cotton varieties" was published. Thus, a methodological guide for multiplication of new promising cotton varieties was born.

Until then, the preliminary multiplication of such varieties was carried out according to schemes approximating the seed production of released varieties.

The lack of a unified methodology for elite seed production work with new varieties gave rise to different approaches to the issues of their finalization by a set of traits. Thus, some agronomists - seed breeders took the shape of boll as the main morphological trait, others - type of branching, others - obliquity, leaf shape and color, and others - pubescence of stem. A similar approach was observed for economic traits. Fiber quality with progeny testing was not controlled in all farms, which had a negative effect when the variety was released on large areas.

The instruction defined a common approach for all seed growers to multiply new cotton varieties. A three-year seed reproduction scheme was established, which includes the establishment of nurseries for first-year family trials (first-year seed nursery), second-year family trials (second-year seed nursery) and multiplication (seed multiplication).

The transition of a promising variety from a two-year to a three-year scheme of elite reproduction should have contributed to improving the efficiency of primary seed production of new cotton varieties.

For unpromising varieties, the two-year seed reproduction scheme remains in place. It is allowed that a nursery family of the first year is rejected if it contains more than three atypical plants.Despite the value of this instruction and its importance for the orderly propagation of new varieties, there are many comments on its individual points.

The current instruction on production of such seeds is morally outdated, it lacks sections defining the need for primary seed production, increasing the role of elite seeds, streamlining the responsibility for varietal conditions of the author of the variety, seed breeder. The norms of relations between the creator of the variety and organizations using this breeding achievement require significant revision.It is envisaged to transfer the seeds of elite new varieties for preliminary multiplication with varietal purity not lower than 96%, which meets the requirements of the standard for the third reproduction.Meanwhile, the main task of the farm of preliminary multiplication of seeds of new cotton varieties is the reproduction of elite seeds and accelerated multiplication of reproductions of new varieties and, if necessary, refinement of new varieties according to varietal quality.

According to the instruction in force since 1986, each farm on new varieties can multiply up to 5 breeding varieties, without spatial isolation.

The analysis of farms' work on preliminary multiplication of new cotton varieties for 2008-2015 shows that not all farms comply with this provision. Some farms sow more than 5 varieties in some years, such as Izbaskensoe, Kasbiy, Uychyn (Kyzyl-Ravat), Akkurgan and some others.

The number of tested new varieties increases every year, so in 2008 there were 55 such varieties, and in 2015 - 76. For 8 years, 4 new varieties were released, 1 variety each from Chimbay, Izbaskent, Uychyn, Kumkurgan elite farms.

According to the rules of the State Variety Testing, each new variety should be propagated in the pre-propagation farm for 3 years and is proposed for release and withdrawal from testing. However, there are varieties that have been tested for more than 8 years. These are Unkurgan-1, Kilozhak and Shonch, and some varieties have been tested for more than 4 years. This loads the work of specialists, does not contribute to the preservation and improvement of varietal purity, on the contrary, further contributes to biological mixing and reduction of varietal purity. (Annex 1).

Consider how inheritance of traits and their variability are interrelated.

§3.4 Inheritance of traits and their variability

It is known that genetic science distinguishes two groups of traits, qualitative and quantitative. At the suggestion of K. Mauser, the concept of essential differences between the "main genes" influencing the manifestation of qualitative traits and the action of polygenes responsible for the inheritance of quantitative traits was introduced. These systems of genes, according to Mauser, have different localization in the substance of the chromosome. Moreover, qualitative traits, as a rule, are inherited monofactorially and are determined by the number of genes, not exceeding the ploidy of the organism. Quantitative traits are inherited by a large number of unambiguously acting polymeric genes. And the more complex the inheritance of a trait, the greater the number of genes that control it.

The study of inheritance of the trait "branching type" of cotton leads to contradictory conclusions. Thus, in the studies (Musaeva D et al., 1983) it is noted that all subtypes (1, 11, 111) of the indeterminate type of sympodium in relation to the limit type behave as obvious dominants, and a 3:1 split is observed in F_2 . At the same time, crossing lines, for example, with 1 and 111 types of fruit branches, leads to inheritance as in polymery.

It was also noted that among hybrids F_2 from crossing of marginal type lines with 1 non-marginal type lines there are individuals with 11 and 111 types of fruit branches, i.e. different from both parents. Such facts require explanation, as they do not fit into the formula of monomeric inheritance. According to selection efficiency, quantitative traits of cotton are proposed to be divided into 2 groups (Simongulyan N, 1980): the first one - traits controlled by the largest number of polymeric genes and, therefore, the most variable productivity and number of bolls per bush.

The efficiency of selection on them is the lowest, and heterosis is more often observed on them; the second - fiber length, strength and fineness, controlled by a smaller number of genes. They are less variable under the influence of environmental conditions, so their inheritance rates are higher and selection is more effective. Such traits as vegetation period, boll size and weight of 1000 seeds occupy an intermediate position.

It should be noted, however, that the subdivision of traits into qualitative and quantitative is conditional, since there are no fundamental differences between them in inheritance. This is evidenced at least by the following fact. Selection on the traits of raw cotton weight of one boll, height of the main stem, vegetation period, length and fiber yield, seed size, as well as in the case with qualitative traits, is quite highly effective.

This is the basis for assuming that they can also be determined by the main genes providing their manifestation in the offspring. According to our observations, the stabilization of each of the mentioned traits separately was observed from the second year onwards

Varieties can combine different type of branches and different degree of leaf lamina dissection (qualitative traits) with different plant height, boll size, early maturity, length and fiber yield, i.e. the noted qualitative and quantitative traits are rather freely combined with each other. This indicates a similar basis of their determination.

Studies show that improving the efficiency of seed production is not easy. Selection of biotypes that could be useful in the future is hampered by veiling of traits with wide mutational and modification variability. Therefore, at present, there is an urgent need to conduct in-depth fundamental research under strictly controlled conditions in order to better understand the genotypic nature of the source material for breeding work and its potential for phenotypic manifestation of useful traits and their interrelationships.

In addition, for the fullest utilization and manifestation of the potential of existing and newly developed varieties, more attention should be paid to improving the overall level of cropping culture and maximizing the potential of agronomic and related measures.

Based on the above, it seems to us that the concepts of trait and property should be understood ambiguously, as is customary. Obviously, the term "trait" should be understood as all qualitative features of an organism determined by the main genes. This concept should include mainly the elements that ensure the growth, reproduction, and survival of the organism under certain conditions. All variations of a trait (in phylogeny and ontogeny) can be imagined as a result of inherited changes in the properties of genes, expressing their potentialities, provided and realized as a result of combinative, mutational and modification variability. For example, the category of traits, first of all, should include the appearance, which is necessary by virtue of evolutionary conditionality:

-generative organs - bud, flower, box, seed, etc..;

-vegetation period in general, and all transitions on this trait should be understood as inherited changes in the properties of the main genes;

-height of plant growth in general - consider all transitions as inherited changes in the properties of the underlying genes.

The manifestation of such widely variable traits as boll weight, productivity and yield can be visualized as the result of the action of genes responsible for boll emergence.

Their number on the bush is not determined by any special genes, but depends mainly on the level of agrotechnics, specific climatic conditions, season conditions, etc. Otherwise, selections of the most productive plants would keep this property unchanged in repeated generations, which in fact does not happen. And it is not accidental, therefore, that modern varieties differ practically little in this feature from the first domestic varieties.

It is well known that new breeding varieties undergo obligatory state testing. At the same time their morphological, economic characteristics and technological properties of cotton fiber are studied in order to select the best of them for production.

Fiber classification is carried out as follows. For cotton fiber testing, two samples are taken from each indoctrination or trial sample during the ginning process. The finished sample is packaged, labeled and sent to the fiber owner or to a laboratory for analysis. Classification until recently was done visually by organoleptic examination of the sample to determine grade and staple length. Only the micronaire index was determined instrumentally. Since 2001, all analyses must be performed instrumentally on HVI systems.

The results of the analyses are recorded in a classification card, which is called a green card because of the green color of the inscriptions and markings. The completed card is returned to the factory, which in turn passes it on to the farmer. On HVI systems, color (reflectance and degree of yellowness), length, micronaire index, relative breaking load, length irregularity, elongation at fiber break, and non-fiber impurities are determined. The fiber quality gradations in the standards tested by the committee decreased regardless of fiber breaking load and maturity factor. It was concluded that the standards did not reflect a consistent decline in fiber quality as the grades decreased. There is no doubt that this conclusion is still true today.

The values of micronaire index and absolute breaking load are close in numerical terms. Their clear relationship within the same type is determined by the overall strong dependence on fiber thickness. Since there is a discrepancy in linear density between fiber types, the microneur index responds to this.

Absolute breaking load is less sensitive to type change, so the reciprocal relationship between microneedle and absolute breaking load will change on different fiber types.

The yarn quality is clearly influenced by the relative breaking load, the linear fiber density and the staple length. These three parameters are strongly correlated, which reflects the real picture of the properties of the different fiber types. Longer fibers tend to have lower linear density and higher relative breaking load. It will be revealed which of these factors has the main influence by fixing them one by one. For orientation, attention should be given to the mature fiber with the highest relative breaking load. This will significantly change the view of both the system of cotton fiber quality assessment and the targeting of breeding work on new cotton varieties.

The same approach is being followed in global cotton production today. At present, cotton-growing countries should pay maximum attention to the cultivation of cotton with a lower micronaire, i.e. thickness. The main task in seed multiplication is to maintain the economically valuable traits determined by the author during variety zoning. At the same time, special attention should be paid to the preservation of traits during rejection and selection of elite material.

§3.5 Study of scrapping and selection of elite material in elite pre-breeding seed farms

In order to study field and laboratory scraping, experiments on new cotton varieties were laid in Izbaskent elite farm of Andijan region with seeds of Unkurgan-1 variety in the amount of 300 individual selections and 50 families of nursery of 1 year;

b) Uchkupriksom of Fergana oblast with seeds of Unkurgan-4 variety in the amount of 150 individual selections and 15 families of nursery of 1 year.

Before sending the seed to the elite farms, the technological parameters of fiber of individual selections were determined in the laboratory of Sifat. According to the results of the analyses, indo-selections with micronaire below 3.8 and above 4.8 were rejected. Selections with micronaire 3.9-4.7 were used for sowing. During the growing season from May to October, the researchers repeatedly traveled to the mentioned elite farms and participated in field screening, rejection of atypical plants, and selection of the best most typical plants for individual selection. In Izbaskent elite pre-propagation farm besides Unkurgan-1 variety 4 more varieties were propagated: Kelajak, Andijan-37, Andijan-40 and Umid. Data on yield at 1 harvest are given in Table 5.

Table 5

Raw cotton yield of new cotton varieties at 1 harvest

№	Sort	Raw seed cotton		Industrial raw cotton		Total raw cotton	
		kg	c/ha	kg	c/ha	kg	c/ha
1	Kelajak	5000	25,0	1980	9,9	6980	34,9
2	And-37	3120	15,6	3890	19,5	7010	35,1
3	And-40	3130	15,7	3920	19,6	7050	35,3
4	Umid	630	14,0	910	19,8	1540	33,5
5	Unkurgan-1	1080	20,0	840	15,6	1920	35,6

The data of the table show that the highest yield of 1 harvest was in the variety Unkurgan-1 and amounted to 35.6 c/ha, which is almost 2 c/ha higher than that of the variety Umid. After the completion of harvesting raw cotton, the economic-valuable traits of multiplied varieties were determined by test samples, which are shown in Table 4.

Table 6 shows that variety Unkurgan-1 by yield 45.3 c/ha shares 1-3 place with varieties Andijan-37 and Umid, has the highest (40.4%) fiber yield, which is 5% higher than variety Umid, 2.2% higher than variety Andijan-37 and 1.2% higher than variety Kelajak. At the same time, variety Unkurgan-1 had low weight (90 grams) 1000 pieces of seeds, which is 40 grams lower than variety Umid. At the nursery 1-year 85 trial samples and seed-by-seed collections were collected, on which the weight of cotton - raw cotton of one boll (g) fiber length (mm) and fiber yield (%) data are shown in Table 5. The data show that the weight of raw cotton of one boll varies from 5.3 to 6.8 (g), fiber length in bolls from 33.0 to 34.4mm, fiber yield percentage from 38 to 42.4%.

Table 6

Economic and valuable traits of cotton varieties

№	Sort	Area	Weight of one box, g	Fiber length, mm	Fiber yield, %	Weight of 1000 pcs	Total raw cotton	Yield
1	Kelajak	1,8	6,2	34,9	39,2	116	7450	41,4
2	Andijan-37	1,8	5,4	34,5	38,2	110	8150	45,3
3	Andijan-40	1,8	5,4	33,5	36,7	112	7155	39,8
4	Umid	0,40	6,1	33,3	35,4	130	1810	45,3
5	Unkurgan-1	0,32	5,5	33,8	40,4	90	1320	45,3

According to the data of laboratory analyses, families were rejected. In 85 families 400 individual samples were collected, or on average almost 5 indo samples were collected from one family. The weight of raw cotton and seeds, fiber length were determined from the individual samples collected. Seed weight varies from 31 to 43 grams and fiber length from 32.4 to 34.4 mm.

Table 7

Summary sheet of laboratory analysis results for the variety Unkurgan-1 in the Izbaskan Elite farm

Current year family numbers	Weight of raw cotton of one box, gr	Fiber length	Fiber yield percentage	Harvest of raw seed cotton			% plant thinning	Seed weight of seed of seed collections kg
				From 1 plant, gr	From 1 row of gr	Ts-ga		
2	6,4	33,8	39,2	37,0	1666	18,5	10	570
4	6,4	33,1	36,0	52,3	1937	21,5	26	540
5	6,5	33,6	39,3	35,5	1692	18,8	4	570
6	6,5	33,2	40,0	48,2	1832	20,4	24	560
7	6,3	33,0	41,7	39,1	1525	16,9	22	560
8	6,8	33,0	41,0	38,3	1838	20,4	4	570
9	6.1	33,2	39,6	44,5	1783	19,8	20	540
10	5,8	33,5	38,0	37,6	1615	17,9	14	500
11	6,7	33,6	41,3	39,9	1834	20,4	8	520
12	6,3	34,4	40,7	48,6	2042	22,7	16	580
13	6,5	33,6	40,0	48,8	2098	23,3	14	590
14	6,4	33,9	42,0	37,4	1682	18,7	10	620
18	5,6	33,4	38,0	44,4	1732	19,2	22	580
19	5,7	33,5	39,2	40,0	1680	15,3	16	530
20	5,8	33,9	40,4	51,8	1760	19,6	32	500
21	5,4	33,9	40,0	39,8	1791	19,9	10	480
22	5,9	34,2	38,8	42,2	1689	18,8	20	490
23	5,9	33,6	39,2	41,8	1727	19,2	16	500
24	5,4	34,0	38,4	30,9	1300	14,4	16	470

25	5,9	34,2	42,0	32,9	1483	16,5	10	470
26	5,8	33,7	44,0	34,9	1605	17,8	8	520
27	5,7	33,7	42,8	37,6	1656	18,4	12	460
28	5,8	33,8	41,2	44,9	1888	20,9	16	490
29	5,5	33,4	38,4	49,5	2029	22,5	18	480
30	5,9	34,1	39,2	36,8	1623	18,0	12	490
31	6,1	33,4	39,3	31,2	1377	15,3	12	540
33	5,5	33,4	39,6	45,2	1807	20,0	20	530
35	5,9	33,4	38,0	36,5	1641	18,2	10	500
36	6,2	33,6	39,7	36,2	1451	16,1	20	490
38	5,8	33,4	40,4	40,4	1778	19,8	12	550
39	5,9	33,5	41,2	37,6	1733	19,3	8	440
40	5,7	33,4	37,6	34,6	1452	16,1	16	480
41	5,7	33,7	39,2	37,6	1618	18,0	14	520
42	6,3	33,1	41,3	36,7	1834	20,4	0	570
43	5,8	33,2	40,0	44,0	1673	18,6	24	510
44	5,6	33,9	40,0	44,5	1694	18,8	24	500
45	5,9	33,4	38,0	33,9	1494	16,6	12	560
46	5,8	33,7	38,4	37,3	1791	18,8	4	540
47	5,9	33,4	39,2	33,7	1595	17,6	6	520
48	6,5	33,2	40,0	44,1	1454	16,2	34	510
49	6,2	33,4	39,3	35,1	1616	17,9	8	570
50	5,7	33,7	40,4	37,2	1636	18,2	12	470
51	6,7	33,3	40,3	37,1	1782	19,8	4	530
52	6,3	33,2	40,0	35,6	1424	16,6	16	500
53	5,7	33,7	39,6	38,8	1473	16,4	24	500
54	6,1	34,1	38,4	33,5	1476	16,4	12	500
55	5,4	34,2	37,6	36,9	1405	15,6	24	450
56	6,5	33,9	41,7	41,8	1590	17,7	24	500
57	5,9	33,4	39,2	38,8	1669	18,5	14	570
58	6,5	33,3	40,0	37,3	1604	17,8	14	580
59	5,6	33,5	40,4	34,0	1564	17,3	8	550
60	6,0	33,9	41,2	34,4	1448	16,1	16	490
61	5,7	33,5	43,2	41,4	1534	17,0	26	550
62	6,3	33,7	39,3	44,1	1812	20,1	18	660
67	6,3	33,8	39,7	44,9	1752	19,5	22	570
68	5,6	33,5	39,2	43,4	1954	21,7	10	560

69	5,6	33,6	39,6	40,3	1652	18,3	18	520
70	6,3	33,8	39,0	48,2	1782	19,8	26	620
71	5,8	33,7	38,4	41,2	16,47	18,3	20	570
72	5,9	33,6	38,8	40,6	1746	19,4	14	580
73	5,6	33,7	42,4	38,6	1606	17,8	14	530
74	5,4	33,6	42,0	35,6	1494	16,6	16	540
76	5,4	33,4	40,4	40,2	1689	18,8	16	550
77	6,1	34,0	39,2	34,7	1424	15,8	18	540
78	5,9	34,0	40,0	48,6	1800	20,0	26	550
79	5,6	33,8	39,6	38,6	1700	18,9	12	580
80	5,4	33,8	38,8	36,5	1572	17,5	14	570
81	5,4	33,7	37,6	45,8	1365	15,2	40	480
82	5,5	34,0	38,4	34,3	1439	16,0	16	530
83	5,5	33,6	40,4	43,5	1740	19,3	20	570
84	5,3	33,6	42,0	49,0	1617	17,9	34	610
85	5,7	33,3	39,6	39,5	1779	19,8	10	660

Selection in primary seed production is used to distinguish heritable genetic variation from modifying, i.e. non-heritable variation, to detect and remove the former and to retain the latter. Soil and other micro-differences in any nursery create modification variability, as a result of which many plants move from one variation class to another in productivity the following year under conditions of cultivation that have changed for them. Most of the plants that have reduced productivity in the selection nursery restore it in the seed nursery. Therefore, in further studies, the reliability of family evaluation during field screenings was analyzed.

The methods of reproduction of original seeds, carrying out selections and technique of their testing in the world practice differ significantly from those adopted in Uzbekistan. On this basis, the rationale for the preparation of original (super-elite) seeds was reflected in our further studies.

CHAPTER IV. A COMPARATIVE STUDY OF THE RELIABILITY OF FAMILY EVALUATION IN SEED NURSERIES BY FIELD SCREENING AND LABORATORY ANALYSIS.

§4.1. study of cotton seed reproduction of new varieties

In 2009, 12 lines were sown in the fields of the Republican Station of Primary Seed Production and Seed Science of agricultural crops from which 1100 individual selections were collected and 60 seed collections were analyzed and proginized on 10 saw gin. Based on fiber length and seed weight data, 350 individual selections were rejected and the fiber of the remaining 850 indo selections were sent to RC Sifat laboratory for quality determination. After receiving the results of the analysis of technological properties of fiber of individual selections made on HVI line, the indo selections with micronaire below 4.0 and above 4.8 were rejected. Prior to sowing for all lines remaining for multiplication, seed samples were submitted to the seed science laboratory for determination of sowing qualities. (Table 8).

Table 8

Sowing qualities of seeds of different lines

№	Options	Seed maturity	Germination energy, %	Seed germination, %	Mechanical damage,%	Weight 1000 seeds, gr	Residual fiber content of seeds, %
1	Line-7	96/8	92,0	96,0	1,7	118	1,2
2	Line-8	97/17	89,0	94,0	1,9	112	1,4
3	Line-9	94/18	91,0	93,0	1,6	110	1,1
4	Line-10	98/5	93,0	96,0	2,1	116	1,2
5	Line-11	98/9	92,0	96,0	1,3	107	1,0

6	Line-12	94/16	89,0	93,0	1,8	119	1,2

As the data of the table show, seed germination of all lines was conditioned, i.e. above 90%. Due to the low quality of gin work, seed samples of all lines had residual fiber content above the norms allowed by the standard (from 1.0 to 1.4%) with the norm of 0.8%.

The highest mass of 1000 pieces of seeds showed lines 12 and 7 respectively 119 and 118 grams, the lowest mass of 1000 pieces of seeds was in line 11 it was equal to 107 grams.

On April 15-16, the field was divided into plots, and in the period from April 17 to 18, sowing of prepared, for reproduction and further development, selection material was carried out. On the variety Unkurgan-1 sown 100 individual selections 30 well plots on the scheme 60h30-1. On the variety Unkurgan-2 sown 64 individual selections and 40 seed lots. In addition, lines 7,8,9,10,11,12 harvested from the 2008 crop were sown in the following quantity after rejection.

Line-7 indosets 90 pieces of seedlings 12 rows; line-8 seedlings 16 rows; line-9 indosets 89 pieces, seedlings 16 rows; line-10 indosets 80 pieces; line-11 indosets 19 pieces, seedlings 48 rows; line-12 seedlings-42 rows.

In order to obtain full-fledged sprouts, feeding watering was carried out on April 27-28, and from May 3 to May 8 the record of sprouts appearance was made. Field germination of seeds of lines used for sowing was determined. Seeds of 7 and 10 lines showed the best germination. The density of standing plants was determined, on June 1 and July 1, the best indicators are lines 7, 10, and 11. On July 1, they had a height of 56 to 59 cm and about 1 boll. The beginning of mass flowering was noted on June 30, at this time began to clean lines from varietal impurities. Subsequently, the growth and development of plants were recorded. The best indicators were in lines 7, 10, 11, which were ahead of other lines in plant growth by 2-5 cm, in the number of sympodial branches by 0.7-1.5 pcs, in the number of bolls by 0.8-1.2 pcs. In all lines, plants were cleaned from impurities. Pure-grade material was found in lines 7-10. Low-grade material was found in lines 8, 9 and 12. The material was completely rejected and raw cotton from these lines was collected and handed over as technical. From the rest of the seed material collected 2000 individual selections more than 800 kg of seed gleanings and 40 samples, including line 7 i.e. 425 pcs of samples 25 pcs of seed gleanings, weighing 525 kg, line -10 i.e. 649 pcs of samples 15 and 276 kg of seed gleanings, the rest of the indo-samples in the amount of 926 pcs collected on varieties Unkurgan 1 and 2, Unkurgan 4.

Yield accounting was carried out as of November 1. The highest yield was obtained from plants of line 7, which was 38.7 c/ha and line 10-38.3 c/ha. The raw cotton weight of individual selections was determined in the laboratory, which ranged from 45 to 72 grams. 420 individual selections were ginned and the fiber obtained was sent to Sifat laboratory. Phenological observations on growth development and yield of plants of different cotton lines are shown in Table 9.

Table 9

Information on phenological observations on growth, development of porosity and yield of plants of different cotton lines

№	Name lines	Field germination, %	As of July 1.			As of Aug. 1.			On September 1.				Number of preserved plants, thousand pcs. ha.	Yield as of November 1, c/ha
			Height, cm	Number of sympodia, pcs	Number of boxes, pcs	Height, cm	Number of sympodia, pcs	Number of boxes, pcs	Height, cm	Number of sympodia, pcs	Number of boxes, pcs	Incidence, % Wiltom		
1	Line-7	74	58	12	0,7	100	15,4	95	115	16,5	12,8	2,4	76,2	38,7
2	Line-8	69	51	11,4	0,1	95	14,8	8,0	112	16,7	10,9	4,7	84,6	36,9
3	Line-9	71	54	11,5	0	97	14,7	7,5	113	16,7	10,1	4,9	84,4	34,1
4	Line-10	78	59	12,0	0,6	103	12,8	10,1	117	16,4	14,6	2,0	75,8	38,3

5	Line-11	72	56	12,1	0,3	103	16,0	9,8	116	17,4	12,7	2,7	74,1	37,7
6	Line-12	70	56	11,6	0	98	16,2	75	108	17,8	10,6	5,1	83,7	35,4

§4.2 Investigating the reliability of 1 and 2 year nursery family evaluations in field screenings and laboratory analyses of released and new cotton varieties

Plant productivity is the main economically valuable trait in cotton. At the same time, productivity is a complex trait, which is determined by the number of bolls and their size, seed weight and early maturity. When selecting for plant productivity, seed growers should keep in mind that most of the economically valuable traits in cotton are mutually related. Usually, the interrelation of traits is judged by pair correlation coefficients. Table 10 shows the correlation coefficients between the plant productivity traits number of opened bolls, total number of bolls, boll size, 1000 seed weight and earliness in Unkurgan-2 and Unkurgan-3 varieties developed in RSPCC. The table shows that there is a close positive relationship between productivity and the number of opened bolls and total number of bolls.

Capsule size and 1000 seed weight have a weak relationship with plant productivity. The correlation between plant productivity and early maturity was negative. In the obtained pair correlations, the influence of each trait separately on plant productivity is distorted by the indirect influence of other factors. To determine the private influence of each trait on plant productivity, we applied beta coefficient. Beta coefficients were also used to judge the efficiency of selection. The arithmetic averages and standard deviations of traits were determined from the data of samples of two varieties (Table 11). Further calculations were carried out using the program "Statistics", where the module of multiple regression analysis allows to calculate beta coefficients (Table 12,13).

Table 10

Correlation coefficients between plant productivity and traits

	Number of opened bolls	Number of boxes	Capsule size	Weight of 1000 seeds	Accelerated maturity
Unkurgan-2 variety					
Productivity	0,795	0,867	0,376	0,392	-0,401

Unkurgan-3 variety					
Productivity	0,808	0,834	0,298	0,166	-0,371

Table 11

Arithmetic averages and standard deviations of traits

	Productivity	Number of boxes	Capsule size, g	Weight of 1000 seeds, g	Ripeness, days	Number of opened bolls
	Unkurgan-2 variety					
Average	63,00	12,21	5,51	112,64	108,40	10,22
St. deviation	20	3,82	0,85	18,76	2,46	3,20
Unkurgan-3 variety						
Average	65,6	13,5	5,4	109,8	112,3	10,7
St. deviation	26,54	6,25	0,92	13,13	3,27	4,26

Table 12

beta coefficients

	Number of opened bolls	Number of boxes	Capsule size	Weight of 1000 seeds	Accelerated maturity
Unkurgan-2 variety					
Veta	0,267	0,675	0,439	0,044	0,007
Errors	0,030	0,031	0,257	0,277	0,015
Unkurgan-3 variety					
Veta	0,303	0,652	0,529	-0,070	0,014
Errors	0,053	0,052	0,040	0,039	0,025

Considering arithmetic and standard deviations and beta coefficients, multiple correlation equations were obtained in standardized form:

For Unkurgan-2 variety

$$\frac{X1-63,0}{20,0}=0,267\frac{X2-10,2}{3,2}+0,675\frac{X3-12,2}{2,46}+0,439\frac{X4-5,5}{3,82}+0,044\frac{X5-112,6}{0,85}+0,007\frac{X6-108,4}{18,76}$$

For Unkurgan-3 variety

$$\frac{X1-65,6}{20,0}=0,303\frac{X2-10,7}{4,26}+0,652\frac{X3-13,5}{6,25}+0529\frac{X4-5,4}{6,25}-0,007\frac{X5-109,8}{0,92}+0,014\frac{X6-112,3}{13,13}$$

3,27

Where X1 - plant productivity, X2 - number of opened bolls, X3 - number of bolls, X4 - boll size, X5 - mass of 1000 seeds, X6 - precocity.

The data in Tables 9-11 show that there is a significant difference between the pairwise correlation coefficients and beta coefficients. This difference is explained by the fact that during calculation pairwise correlations are distorted by the indirect influence of other traits. The existence of a relationship between productivity and its constituent traits can be judged more correctly by beta coefficients.

Beta coefficients are comparable measures of the private effect of each trait on the plant productivity trait and are indicators of selection efficiency. The prediction of selection efficiency is carried out as follows.

According to the equation of multiple correlation in standardized form, a change by one standard deviation of the number of opened bolls $\sigma 2 = 3{,}2$ in the variety Un-Kurgan-2 leads to a change in plant productivity by $\beta 2 = 0{,}267$ standard deviation $\sigma 1$ or by $\beta 2 \sigma 1$ =0.267*20.0=5.34. Changes in productivity at selection by one standard deviation of other traits are shown in Table 13.

To increase productivity by selection, trait variation should be directed in a positive direction if the coefficients β are positive, and in a negative direction if the coefficients β are negative.

Table 13

Number of opened bolls	Number of boxes	Capsule size	Weight of 1000 seeds	Accelerated maturity
Unkurgan-2 variety				
5,34	13,50	8,78	0,88	0,14
Unkurgan-3 variety				
8,04	17,30	14,04	-1,86	0,37

Thus, the coefficients β determine the degree of dependence of plant productivity on each of the traits of its components and allow predicting the efficiency of selection. The conducted research allows us to draw the following conclusions:

1. Under the current method of reproduction of new cotton varieties, the authors of the variety practically do not participate in the selection of original seeds and often transfer for multiplication the source material with varietal purity corresponding to the quality of the 3rd reproduction.

2. When conducting field screenings, it is necessary to make fuller use of modification deviations in productivity, but it is necessary to rigidly cull hereditary changes, removing underdeveloped, diseased and reduced forms.

3. To determine the fiber quality of individual selections and trial samples of seed collections of new breeding varieties in pre-propagation farms on the territory of the Republic of Uzbekistan should be analyzed on the HVI line.

4. Indo-selections of nursery families should be rejected with fiber quality by micronaire below 4.0 and above 4.8, which will contribute to the use of quality elite material for sowing.

§4.3 Assessing the quality of field screening and scraping of families

.... Seed breeding is one of the ways to manage variety adaptability based on the ecological principle of creation and existence of variety genotype. It should be considered as the most effective and organizationally available means of intensification, biologization and ecologization of processes in crop production on the basis of selection of seeds with high yielding properties due to epigenetic processes of the previous ontogenesis. In this case, the process of formation of yield properties of seeds, providing the realization of agro-ecological conditions, includes both adequate replacement of technogenic means by biological ones, and more rational use of natural and technogenic resources (ecological seed production), thus providing energy efficiency on the basis of exogenous impact on the genotype through the phenotype of the variety, sustainability, environmental protection and profitability of agricultural production as a whole. This can be done purely technologically in the process of cultivation of the variety and seed formation on the mother plant, as well as on the basis of selection of high-yielding batches of varietal seeds according to their yield properties when growing them both in different agrotechnical and egroecological conditions.

According to the data of annual reports of elite-seed farms, we analyzed the quality of field browse for 2000-2011 for varieties AN-Bayaut-2 and C-6524 (Table 14).

The tables show that a low percentage of family rejection by atypicality is allowed in both nurseries. The main percentage of family rejects is due to agrotechnics. Untimely implementation of agrotechnical measures in the crops of seed nurseries leads to a large rejection of families by desiccation, overgrowth, and thinning.

All these indicators taken together give a large scope of the reason for their rejection. The calculations and the conducted analysis on comparison of indicators of total rejection of families show that the actual criterion of materiality is equal to t_{Φ} =4,41 . The same difference is confirmed by the smallest significant difference: NDS_{05} =9,718 with the difference of the compared variants equal to . $\Delta = 18{,}96$

By atypicality index of rejected families on varieties AN-Bayaut-2 and C-6524 show significant difference of their differences: t_{Φ} =8.98 t_{05} =2.26 with the least significant difference NDS_{05} =2.44; while the difference is $\Delta = 9{,}71$.

Table 14

Analyzing the quality of field screenings in elite seed production facilities farms on An Bayaut-2 variety

Years	Nu	First year nursery		Second year nursery	
		Total	Rejected families in %	Total	Rejected families in %

			Total	By atypicality	By agronomy	Lab tests		total	By atypicality	By agronomy	Lab tests
2000	3	3192	46,1	2,2	27,8	16,1	1012	25,6	4,0	9,1	12,5
2001	2	3228	55,5	2,6	35,0	17,9	1600	32,5	3,9	15,5	13,1
2002	2	2572	45,4	2,3	31,7	11,4	920	26,1	6,3	9,4	10,4
2003	3	3273	33,2	3,1	19,0	11,1	2409	35,6	3,3	25,0	7,3
2004	3	3339	56,0	0,6	37,3	18,1	1635	39,2	3,6	25,2	10,4
2005	3	2000	48,9	0,2	24,7	24,0	738	30,8	0,8	23,3	6,7
2006	2	2700	47,6	0,9	31,9	14,8	2784	41,9	1,4	19,3	21,2
2007	3	2700	35,7	0,6	24,6	10,5	1197	42,1	3,0	32,2	6,9
2008	3	2682	44,7	0,9	23,7	20,1	1308	39,0	2,2	27,7	9,1
2009	3	3060	39,7	0,3	26,1	12,8	1253	27,0	0,2	14,2	12,6
2010	3	3027	50,5	6,0	25,8	18,7	1636	45,1	3,0	29,4	12,7
2011	3	3424	53,5	2,0	29,5	22,0	1455	35,0	1,2	22,0	11,8
Average		1087	47,7	2,2	27,0	18,5	495	35,9	2,5	20,9	12,5

The rejected families according to agrotechnical indicators do not give the significance of their difference both by the criterion "t_Φ " and by the smallest significant difference: t_Φ =1,95, and t_{05} =2,26, NDS_{05} =9,64 at the difference of their difference $\Delta = 8{,}31$. The same position of rejection also by laboratory analysis: t_Φ =0.404, and t_{05} =2.26 and NDS_{05} =5.146 for the difference of the compared variants . $\Delta = 0{,}92$

Calculations of determining the relationship between the volume of sown families with the total rejection of variety AN-Bayaut-2 give a low correlation relationship r=0.25, and variety C-6524 r=-0.05, ie rejection practically does not depend on the volume of sown families.

At the same time, the confidence intervals of total rejection of families of varieties AN-Bayaut-2 and C-6524 do not include some indicators:

- for the AN-Bayaut-2 variety, the confidence interval is $D = \overline{3} \pm t_{05}\ S_x = 66{,}9 \pm 2{,}26 \times 1{,}988 = (62{,}4 - 71{,}4)$ which does not include 59.4; 80.1; 62.0;

- for grade C-6524 the confidence interval is

$D = \overline{3} \pm t_{05}\ S_x = 48{,}99 \pm 2{,}26 \times 3{,}282 = 48{,}99 \div 7{,}419$ which does not include 36.5; 41.2; 62.8; 68.7.

Checking the accuracy of the obtained indicators of total rejection for variety AN-Bayout-2 shows that the doubtful indicators 3_1 =59.4 and 3_2 =80.1 are within the range of possible random fluctuations, which should not be excluded when calculating the average.

The same situation corresponds to the rejection rates for variety C-6524: the questionable rates 3_1 =33.2 and 3_2 =63.7 are also within the range of possible random fluctuations.

From the analysis of statistical data of family rejection rates it follows that the used rejection methodology does not provide an opportunity to correctly determine (estimate) the probability of rejection and requires the development of a better rejection method.

Table 15

ANALIZ

quality of field browse according to annual reports in elite seed farms for variety C-6524

Years	Number of elite farms	Nursery of 1 year					Nursery 2 years				
		Total seeded families, pcs	Families rejected, %				Total seeded families, pcs	Families rejected, %			
			Total	By atypicality	By agronomy	On the lab tests		Total	By atypicality	By agronomy	On the lab tests
2002	**1**	**1500**	**63,3**	**9,1**	**37,6**	**16,6**	**450**	**44,4**	**9,0**	**24,3**	**11,1**
2003	**2**	**2320**	**59,4**	**10,5**	**23,0**	**25,9**	**646**	**36,5**	**7,5**	**20,6**	**8,4**
2004	**2**	**2260**	**80,1**	**12,7**	**51,2**	**16,2**	**926**	**68,7**	**10,4**	**50,3**	**8,0**
2005	**3**	**4100**	**62,0**	**11,3**	**29,3**	**21,4**	**1350**	**52,3**	**10,9**	**27,9**	**13,5**
2006	**2**	**3000**	**63,6**	**12,1**	**39,7**	**11,8**	**900**	**42,0**	**9,7**	**25,6**	**6,7**

2007	**2**	**3000**	**70,0**	**13,2**	**37,6**	**19,2**	**900**	**44,4**	**10,1**	**26,0**	**8,3**
2008	**2**	**2900**	**65,7**	**10,8**	**33,7**	**21,2**	**868**	**41,2**	**12,8**	**18,0**	**10,4**
2009	**3**	**3360**	**70,7**	**14,4**	**50,3**	**6,0**	**1250**	**62,8**	**13,6**	**47,8**	**1,4**
2010	**3**	**4500**	**62,0**	**12,5**	**32,0**	**17,5**	**1270**	**42,8**	**11,9**	**26,8**	**4,1**
2011	**3**	**4420**	**72,2**	**11,9**	**42,6**	**17,7**	**1674**	**54,8**	**12,5**	**36,8**	**5,5**
On average per 1 el.hozvo		**1363**	**66,9**	**11,9**	**37,7**	**17,3**	**445**	**49,0**	**10,8**	**30,4**	**7,8**

In the nursery of the 1st year, rejection amounted to almost 67% of the number of sown families, including about 12% for atypicality and almost 38% of families rejected for agrotechnical reasons. Large agrotechnical defects of more than 30 % were also observed in the nursery of 2 years. According to the current methodology it is necessary to sow not less than 1500 families in the nursery of 1 year, and in the nursery of 2 years not less than 400 families of the nursery of 1 year and thus the methodology assumes to carry out rejection of 73% of families of the nursery of 1 year and almost 40% of families of the nursery of 2 years.

This situation leads to the workload of workers of elite farms conducting field reviews and does not allow to qualitatively conduct field scraping, putting the main emphasis on scraping on agrotechnics.

The instructions for the production of elite and first reproduction seeds of the released cotton varieties do not provide for the elite-seed production taking into account the genetic characteristics of varieties, methods of their breeding and selection, duration of their presence in production. After all, the character of variability of morphological and economically valuable traits and their stability should not be identical in varieties with different origin and genetic potential (mutant, interspecific hybrid, intermutant hybrid, etc.). However, rejection on agrotechnics does not show the low quality of families on varietal purity and economically valuable traits, but indicates the low level and timeliness of agrotechnical measures on elite crops. At the same time, families with plants with high potential for further multiplication may be culled. Obtaining pure seed is a problem that involves not only varietal traits, breeding and seed production methods, and the influence of different developmental conditions, but also the precautions that are necessary when harvesting, ginning and transferring seed.

An important condition for preserving the varietal qualities of elite material is the reliable evaluation of individual selections intended for collection.

§4.4 Assessment of criteria for collection and rejection of individual selections

Individual selection of plants for cotton variety renewal in elite farms should be carried out in two steps. Preliminary evaluation and allocation of bushes for selection are provided after the first normally opened bolls appear, and final evaluation - just before harvest. Such an order was introduced in elite farms due to the fact that by the harvest period the main morphological features of the cotton plant undergo significant changes, and it is not possible to determine the shape of the boll at all. In addition, a thorough selection in one go just before harvesting is difficult because this operation has to be carried out in a very short time. In such conditions, selection of bushes atypical for the variety is not excluded, and in subsequent years, the labor intensity of cleaning elite-seed crops from impurities increases.

Despite the obvious advantage of double plant evaluation in elite farms, individual selection is mostly done in one go - often just before harvest.

It is known that each indo-selection differs even within the same bush, boll, seed. On the lower tier of the cotton bush flowers open, pollen fertilization occurs around June, on the upper tiers 1-2 places 7-8 sympodia in July-August.

June flower opening period after fertilization to maturation, that is, the embryonic period in the physiologically young organism of the mother plant fall in a longer daylight hours, hot dry weather. In the embryonic period conditions change within one bush the shape of the boll, seed, technological qualities of fiber and other indicators. The offspring of such plants from year to year lose varietal purity, i.e. genetic homogeneity, initial economically valuable features and fiber quality. To clarify the criteria of plant evaluation during individual selection, we sowed two samples of the main regionalized variety C-6524 of Yukori-Chirchik and Chinaz elite farms. To restore seed sowing qualities and clean the variety from impurities, a sample of the original elite was sown in 2009-2011.

The beginning of flowering, the date of opening of the first boll and fruiting were determined for all plants in the experiment. At the beginning of harvest ripening, 200 individual selections were identified in each of the two experimental plots, including 100 with the highest fruiting regardless of their maturity (by the beginning of boll opening) and 100 selections with the earliest opening of the first boll and fruiting not lower than the average level.

Individual selections were subdivided by raw weight into classes and for each class average data on the main indicators characterizing plant development and yield formation were determined. No significant difference in trait correlation was found between the selections similar in the method of evaluation in the reproductions of the initial and last elites. The difference in trait correlation was mainly determined by the principle of plant evaluation.

The distribution of individual selections by weight of home-grown raw cotton in all four groups had a single-vertical curve of the variation series. This indicates comparative equalization of the variety materials and normal growing conditions in the experiment.

When selecting plants for yield - accumulation of fruiting organs, and irrespective of the timing of opening of the first boll - the highest number of individual selections (mode) was in the class of 70 g ,
and when selecting for early opening of the first bolls and fruiting not below the average level, it was more often in the range of 50 g.

When selecting plants according to their fruiting, a certain correlation was observed between the weight of raw seed cotton of individual selections and the dates of opening of the first flower, first boll, boll opening rate, total fruiting and boll size (Table). Evaluation of plants with high fruiting not only by boll opening rate in September, but also by the onset of the ripening and even flowering phase allows to identify more complete individual selections.

In case of selection of plants only on the basis of the earliest opening of the first boll, the correlation of raw seed yield with other economically valuable traits was much less pronounced or was not traced at all. Variation of indices for individual traits within yield classes was less than in case of selection based on fruiting (Table 16).

Table 16

Plant development, yields at individual selection for fruiting and early maturity

Selection classes by raw weight	Number of cases	Class averages							
		Fruiting	Flowering start date	Date of maturity of the first boll	Pace of discovery boxes on			Packs of collected boxes	Box weight (g)
					10/I X	20/I X	30/I X		
Selection by highest fruiting									
20(16-25)	2	13.7	8.07	3.09	0	2.7	7.0	4.8	4.1
30(26-35)	3	16.8	9.07	1.09	0.1	3.0	9.8	6.5	5.8
40(36-45)	7	13.0	30.06	25.08	3.0	7.8	10.2	6.7	6.4
50(46-55)	12	14.0	30.06	26.08	1.7	5.4	9.6	7.7	6.3
60(56-65)	17	14.5	28.06	25.08	2.3.	6.4	10.4	8.7	6.6
70(66-75)	23	14.0	27.06	24.08	3.0	8.8	11.7	9.8	6.6

80(76-85)	12	14.0	25.06	22.08	3.2	7.6	11.8	10.4	6.7
90(86-95)	12	18.5	25.06	22.08	3.7	8.3	13.2	10.9	6.9
100(96-105)	4	19.0	27.06	23.08	3.0	8.5	13.4	11.2	6.8
110(106-115)	5	20.8	23.06	20.08	4.1	10.8	15.8	12.0	6.9
120(116-125)	1	23.0	22.06	19.08	4.8	10.9	16.3	15.2	7.0
130(126-135)	2	20.0	21.06	19.08	5.1	11.4	17.4	16.0	6.8
Selection by early opening of the first box									
20(16-25)	3	9.5	1.07	22.08	4.1	7.3	10.1	8.5	5.9
30(26-35)	3	9.9	2.07	23.08	4.0	6.8	9.3	6.7	5.9
40(36-45)	18	10.1	2.07	21.08	4.2	7.1	9.8	7.1	6.0
50(46-55)	34	10.7	2.07	22.08	4.4	8.1	10.2	8.1	6.2
60(56-65)	25	10.7	1.07	20.08	4.5	8.4	10.7	8.4	6.4
70(66-75)	11.9	11.9	30.06	19.08	4.3	10.2	11.0	10.2	6.5
80(76-85)	12.4	12.4	30.06	20.08	3.4	10.7	10.8	10.7	7.0

In plants selected for the earliest opening of the first boll with fruiting not lower than average, there was no definite dependence of raw seed cotton yield on the timing of the first flower and first boll opening. At seed cotton yields from 20 to 80 g per bush, the difference in the onset of flowering was within ± 1 day by class, and the difference in the onset of boll opening ± 2 days. Consequently, individual selection of average fruiting plants by the beginning of boll opening, and even more so, selection without taking into account the total fruiting of bushes does not guarantee a full yield of raw seed cotton.

When evaluating plants for individual selection at one time before harvest, seed growers at elite farms often focus on the number of bolls that have opened by the time of screening. Therefore, with such a principle of selecting plants for selection, viewing them earlier in the calendar will be associated with an increase in the number of low-weight individual selections. After all, the character of variability of morphological and economically valuable traits and their stability should not be identical in varieties with different origins and genetic potential (mutant, interspecific hybrid, intermutant hybrid, etc.).

Individual selection, cleaning, rejection of families are carried out by differently qualified seed breeders, who can not always give a correct assessment of the families left, can not recognize the best genotype and as a result of erroneous selection and rejection of families can be a complete loss or sharp deterioration of the original economic-valuable traits and technological qualities of fiber.

The existing instruction states that if a family has more than 2% of deviated plant forms, it is completely rejected, despite the fact that the remaining 98% of plants are productive, homogeneous and can give in the offspring fiber with high technological quality. Then rejection is carried out according to the data of laboratory analyses conducted in elite seed farms: boll size, fiber length, seed weight, and determination of technological properties of fiber - breaking load, linear density and relative breaking load are carried out centrally at the discretion of the Republican Center for Cotton Seed Production and the best families are selected for planting nurseries of the current year.

Evaluation and rejection of elite material is carried out on the basis of a set of features, taking into account climatic and agrotechnical conditions of seed growing. The disadvantages of this method of rejection and selection of elite material for sowing are:

- very low performance level

- not fully reliable assessment of individual selections, which is based only on fiber length and seed weight

- arithmetical errors in the variation series, which can distort the scrap data

In modern conditions for the development of technology of assessment and rejection of elite material, the issues related to the reduction of labor and time costs for calculating variation series, improving the quality and competitiveness of raw cotton in the world market are of paramount importance. In order to achieve normative indicators of raw cotton, it is necessary to improve technologies to improve its quality, taking into account the specific properties of selection varieties of cotton. Qualitative and quantitative indicators of cotton fiber and other by-products largely depend on the level of selection and rejection of material. Therefore, to solve the problem of improving the quality of raw cotton, the problems of improving the technology of selection and rejection play an important role.

The solution to the problem of optimizing the seed production process when stabilizing cotton varieties requires sufficiently homogeneous in size, shape and speed of development individuals with high productivity and resistance to diseases, giving high quality products, etc. The use of directed selection methods does not always solve this problem: hypertrophy of some functions is inevitably associated with deterioration or weakening of others, which is the usual "payment" for selection. This approach can be contrasted with another one, in the framework of which the phenotype that minimally deviates from the average of various varying traits is considered to be the most adapted to the diverse requirements of the environment.

§4.5. Study of plant modal selection methods

Each variety was sown separately in plots of 5 thousand plants. The layout was 60 x 30 x 1. Agrotechnical measures were common in production. To collect the source material plots were conditionally divided into squares of 4 rows with 40 - 50 plants in each. Selecting from the center of the row one plant at a time, 100 plants of Sultan variety and 40 parental generation (P) were selected. Each plant was characterized taking into account morphological and physiological features - a total of 17 measured or counted traits for vegetative organs, such as height (cm), number of fruit branches, height of the first fruit branch (hs) , total number of bolls, number of opened bolls, fallen fruit organs, branching type, etc., were taken into account. All descriptions, counts and measurements were carried out according to generally accepted methods. To characterize the variability of generative organs, two bolls each were selected from the first places of the 3rd and **4th** fruiting branches. Harvesting was individual for all described plants.

The results of field observations and laboratory analyses were subjected to statistical processing: mean values of each trait (X), error of mean value (± m), limits of trait variability, coefficient of variation (C v) and dispersion (∂). The number of lobes in a boll, number of seeds, weight of fiber, seeds and one boll, fiber yield and length were determined in laboratory conditions. Along with the study of variability of all traits, their correlation (r) was assessed.

After processing the results of measurements and observations, three areas of variability were identified for each trait: close to the mean value (M±m) and those more than two standard deviations to the right and left of the distribution center. Plants that fell into the central zone for most traits were referred to the group conventionally labeled "Mo" ("average phenotype"); plants that fell into the right part - "M+"; plants that fell into the left extreme part - "M".

To test plants for progeny from the first and second places of the 3rd and 4th sympodia of each bush, two boxes were taken and the seeds

contained in them were sown in spring 2008 in the same plot where the previous generation was grown, i.e. the conditions of the experiment were equalized in the continuation of the experiment. The sowing scheme was the same as in the previous year. Control seeds from production plants were sown on plots between the compared groups. Seed germination was determined in the field.

The relationship between the traits was plotted in the form of correlation "fields", graphically - in the form of curves. The obtained values of intrapair correlation coefficients are considered to reflect a strong relationship at r> 0.7, average at r=0.3-0.7 and weak at r=0.3. < 0,3.

The variance complexes were used to estimate (for each trait) the degree of hereditary heterogeneity of the selected groups of plants from the data for their progeny by means of the intraclass correlation coefficient (r w). We used the analysis techniques of E.H. Ginzburg

It is important to take into account that in the process of modal selection it is possible to change the structure of a variety in the desired direction. This is also reflected in changes in the nature of relationships between traits. Selection is carried out on the "average" phenotype, the strengthening of correlation in the corresponding group of plants clearly indicates the effectiveness of modal selection, forming a plant with a new, more optimal phenotype. The external expression of any trait depends on both environment and genotype, and even minimal estimates of the contribution of heredity in determining the diversity of the cotton population by the peculiarities of the structure of vegetative organs indicate the significance of such a contribution (Table 17).

According to the data of the Table, we can judge about the correlation between the traits of cotton variety population and its heterogeneity. Consequently, there are reasons to expect certain effects of selection according to the scheme proposed by us.

Table 17

Intraclass correlation coefficients (rw) in the group of plants obtained through modal selection process

Number attribute	X	Se	Smr	S a	r w	F*
1	123,36	443,67	1714,04	100,74	0,185	3,86
2	19,15	11,12	51,37	3,18	0,221	4,58
3	19,93	7,70	48,89	3,27	0,300	6,35

4	104,58	492,39	1489,07	79,07	0,138	3,02
5	5,20	0,42	3,10	0,21	0,335	7,35
6	65,07	621,71	3253,89	208,74	0,251	5,23
7	5,15	0,81	4,96	0,33	0,289	6,14
8		19137,3	84408,4	5176,46	0,210	4,41
9	335,72	7	6	0,06	0,131	2,89
10	6,66	0,35	1,14	28,77	0,214	4,44
11	23,05	105,36	468,07	20,49	0,254	5,28
12	17,60	60,38	318,80	28,25	0,087	2,21
13	77,71	293,75	649,99	23,52	0,238	4,93
14	23,47	75,27	371,85	7,80	0,150	3,22
15	51,17	44,29	142,64	0,06	0,108	2,53
16	2,06	0,49	1,25	2,94	0,107	2,52
17	18,85	24,44	61,51	59,05	0,107	2,51
	63,37	494,65	1239,37			

*Statistically, in all comparisons, Fisher's ratios (F) for the 17 traits are significant at levels of $P \geq 0.001$.

The data of the Table show that modal selection allowed to obtain a cotton population that is practically not inferior to the control group in terms of comparable economic-valuable traits. The modal group is characterized by earlier and more friendly germination not only in comparison with the plants obtained by directional selection, but also in comparison with the control. Germination in the control was 93.3%, in the subgroup M - 96.6; Mo-98.0; M+ - 92.5%

Table 18

Comparison of groups of plants obtained by modal and of directional selection, by precocity

Group and subgroup	Scope of the sample,	Field germination seeds, %	Number of days from sowing before the	Laying height first fruit hs branches	Number of opened bolls by 15.X.%	Percentage of plants 100% disclosed boxes, %
Control 2008 г.	174	82,1±2,9	137,2±1,6	6,64±0,08	70,0±3,5	16,1±2,6
M_group	166	80,3±3,1	133,9±1,7	6,25±0,06	83,3±2,9	37,3±3,4
Moe	357	85,8±1,9	133,6±0,9	6,60±0,04	77,7±2,2	19,3±1,7
M+	213	80,3±2,8	139,6±1,3	6,96±0,05	47,7±3,4	5,2±2,1
Control 2009 г.	152	93,3±2,1	140,4±1,5	6,95±0,10	88,4±2,6	27,0±1,6
M_sub-. group	114	96,6±1,6	132,0±2,1	6,39±0,09	97,8±1,3	78,9±1,6
Moe	644	98,0±0,5	134,7±0,5	6,68±0,04	93,6±0,9	63,2±0,9
M+	130	92,5±2,2	142,9±1,7	7,05±0,05	87,8±2,8	28,5±1,8

At the same time, if we compare the Mo group with the control by such a characteristic as the proportion of plants with 100% open bolls, we will see its obvious advantage: by October 15, there were only 27% of plants with fully opened bolls in the control and 63.2% in the Mo group (Table 18). It is evident from the Table that selection led to a significant decrease in fruit drop in the Mo group and Mo subgroup.

Modal selection also led to a sharp divergence of the compared populations also by wilt resistance, determined at the end of the vegetation period using an oblique cut at the root neck. The degree of damage was assessed on a three-point scale (Table 19).

Plants obtained in the process of modal selection turned out to be more wilt-resistant. This effect is especially clear in F2.

Thus, the modal group is superior or at least no worse than the control population in terms of economic-valuable traits characterizing plant productivity, wilt resistance and suitability for mechanized harvesting. The same trend is evident when comparing the Mo, M+ and M- subgroups. These results are consistent with the general ideas about the optimality of the "average" phenotype in the population and the effectiveness of modal selection in cotton production.

Table 19

Degree of wilt infestation of cotton groups obtained by directional and modal selection

Group and subgroup	Scope of the sample, ex.	Number of diseased plants per 10.X, %			
		mildly affected	average affected	strongly po-wounded	total
Initial material (20 yrs.).	180	8,88	2,77	2,22	13,38±2,5
Control 20 г.	174	7,47	4,02	2,29	13,80±2,6
M_group	166	10,84	6,02	-	16,86±3,0
Moe	357	5,32	1,68	2,24	9,20±1,5
M+	213	4,22	7,51	2,34	14,10±2,4
Control 20 г.	152	3,28	4,60	2,63	10,90±2,4
M_sub-. group	114	7,89	0,87	-	8,76±2,7
Moe	644	2,01	1,55	-	3,56±0,8

M+	139	5,38	2,30	10,76	18,64±3,4

For this purpose it is advisable to improve the technology of rejection, which at lower labor costs, has higher technological performance.

.4.5 Study of plant modal selection methods.

The development and multiplication of new and released cotton varieties in elite-seed farms is carried out by individual selection with subsequent progeny testing. This method is quite suitable for varietally pure varieties with already established population heterogeneity to a greater or lesser extent.

But modern cotton breeding lines are getting
mainly on the basis of distant hybridization, using donors of several species, which leads to obtaining complex hybrids with a long time splitting by a number of morphological and economically valuable traits. Therefore, the used methods of elite-seed breeding work do not always allow to quickly solve the problem of stabilization of the obtained new breeding material on the complex of traits, especially since the use of directed selection on several traits, without taking into account the rest, in the presence of negative correlation inevitably leads to weakening or deterioration of other important for the variety qualities.

We compared the method of selection by correlation of traits, used on the released cotton varieties and on the varieties of preliminary propagation, in combination with stabilizing (modal) selection on the complex of traits of the bush habitus of the plant. The aim was to identify a way to bring a new variety (breeding line) more quickly to the requirements of the elite standard.

The experiment was first laid in 2008. The source material was 240 individual selections of the new variety Sultan. In the first year, 50.4% of families were culled due to atypicality, which indicates high heterogeneity of the variety's source material.

In 2009-2011, three nurseries were sown adjacent in origin: 1- nursery of initial families - (I); 2- nursery - control methodology (A); 3- nursery - modal selection methodology (B).

All seed production activities were carried out in accordance with the "Instruction on production of elite and first reproduction seeds of zoned cotton varieties and "Instruction on preliminary multiplication of seeds of new cotton varieties".

After field screening and clearing of non-rejected families from non-typical plants in the modal nursery, all remaining plants were described by quantitative traits of bush habitus, taking into account: a) plant height, b) height of the first sympodium, c) number of fruit branches, d) total number of bolls, e) number of open bolls. According to the obtained data, variation series were compiled and the average variant of the trait was determined. Families and plants in them, the average data of which corresponded to three of the five listed traits of bush habitus, were selected for further reproduction by the method of modal selection. For economically valuable traits, according to the results of laboratory tests, families and individual selections in them also corresponding to the average of variation series were selected with a tolerance within the double square error of the experiment.

During field inspections, according to the current rules of cotton seed production, families with more than three atypical plants in their composition were culled from further elite reproduction, and the remaining families were cleared of atypical plants.

Table 20

Results of field screenings by nursery of the techniques under study

Year of experience	Methodology selections	Volume of field marriages by non-typicality, %	Volume of field venting from non-tropical plants, households
2009	И	29,4	0,9
	А	29,8	0,4
	Б	21,2	0,9
2010	И	43,1	1,8
	А	40,4	1,6
	Б	31,5	1,5
2011	И	48,0	1,6
	А	21,1	1,3
	Б	19,3	1,3

As can be seen from the given data (Table 20) the volume of field rejection of modal selection families was annually less against the control and at the same time more rapid stabilization of phenotypic homogeneity of the population occurred. Apparently, description of plants by quantitative traits of bush habitus contributes to additional detection of heterozygous plants and faster purification of varieties from them in comparison with the technique of selection of elite materials by trait correlation. In the material of both methods quite often there are plants with raw cotton yield more than 100g, but otherwise in morphological features indistinguishable from the main type of variety. According to the method of selection by trait correlation, such plants are "the best" and are not subject to rejection, while by the method of modal selection they are rejected from further reproduction.

When these plants heterosis in yield were sown in 2009-2010 in a separate experiment (Table 20), the split in morphological traits in both the first and second progeny was significantly reduced, with several new biotypes appearing in the second progeny.

Heterosis plants of the modal method in the first progeny did not give atypical forms for the variety at all, and in the second progeny their number was insignificant. Consequently, selection of the "best" plants in a new variety that has not been reliably tested for varietal purity partially involves heterosis splitting forms in further reproduction, which prevents the stabilization of the variety in terms of morphological traits.

In the process of refining the material of both techniques on varietal purity, some difference of plants on quantitative traits was observed: plants of the modal selection nursery constantly had more by 1-2 fruit branches and, accordingly, a slightly higher number of bolls at increased precocity, which with a relatively equal weight of the boll difference of 0.1-0.2 g, ultimately provides an excess yield of the material of the modal technique over the control.

Table 21

Splitting of progeny of heterosis plants morphologically

Year of experience	Methodology selections	Number of families	General quantity	Including. typical. grade,%	Rejected_forms__ small.:large.:large.:small.:large.:small. kor-k:kor,:kor.:kor.:kor.:kor.:kor. 1.5type:1.5type : 1 type :1 type : II type	Others
2009 - 1st generation	Control Modal	34 26	1643 1166	59,8 100,0	31,9 3,3 0,9 3,5 0,3 0 0 0 0 0	0,3 0
2010 - II generation	Control modal	128 49	3885 1442	56,4 99,7	18,8 5,8 2,2 4,3 1,3 0 0,1 0 0 0,2	11,2 0

According to the data of the first three years of trials (Table 22), the average yield of seed raw cotton of the families in the nursery of the modal selection method was slightly lower than that of the initial families and families of the control method. This phenomenon can be explained by the presence of a rather large number of heterozygous plants in the nurseries of initial families and the control, which are not detected during field observations, but having heterozygous nature, give split progeny in further reproduction. This position is confirmed by the presence of a greater number of plants atypical to the variety in the initial families (I) and families of the control method (A) than in the families of the modal method (B). In addition, it should be borne in mind that the individual weight of raw cotton (and seeds) of the best phenotype individual selections of the control technique was annually 1.5-1.8 times greater than that of the modal individual selections, which inevitably affected the density of plant stand and, ultimately, the yield of families in nurseries.

Table 22

Yield of raw seed cotton

Year of experience	Variant techniques	N	X	S	V	Sx	Sx%
2009	И	101	38,8	8,8	22,6	0,9	2,2
	A	73	39,3	8,8	22,4	1,0	2,5
	Б	81	37,2	8,7	23,3	1,0	2,7
2010	И	181	21,8	4,2	19,3	0,3	1,4
	A	52	21,0	5,3	25,2	0,8	3,8
	Б	61	21,0	4,9	23,3	0,6	2,9
2011	И	178	30,0	6,1	20,3	0,5	0,7
	A	90	29,0	6,7	23,1	0,7	2,4
	Б	92	26,0	6,1	23,5	0,6	2,3

At the same time, the results of testing of elite samples of different years of production (Table 23) allow us to state that at least equal plant density in nurseries already provides more or less the same yield of both the material in general and individual plants (average values) in particular.

Fiber yield is a trait stably inherited and organically related to other economically valuable traits, so it is practically impossible to change its indicators without changing other economically valuable properties of the variety.

Genetically common origin of the elite material of both techniques determines their difference in fiber yield within the error of experience (Table 24). The inverse correlation of fiber length with fiber yield also determines the insignificance of differences in fiber length between the materials of both methods (Table 25). In the last two years, the advantage of the material of the modal technique on fiber yield becomes more obvious, most likely, there is no change of the trait in this material, on the contrary, the increase in fiber length by families of the method of selection by trait correlation is affected, which undoubtedly led to an undesirable decrease in fiber yield by families.

Probably, this is another confirmation of constancy of economically valuable traits of the material of the modal method of selection by correlation of traits with some subjectivity of selection of elite materials by individual traits.

Table 23

Test results of elite samples from different years of production for 2009-2011

Year of production elites	Options	Maturing. 50% days	Sorto-waya purity %	Harvest. 1 collection c/ha	Harvest. on the plant.	Weight box.	Output fibers %	Length fibers mm
2009	И	131	95,8	52,2	77	6,6	36,2	34,3
	A	134	97,8	50,8	78	6,6	37,1	33,8

	Б	132	96,6	49,8	77	6,7	38,2	34,4
2010	И	134	90,9	44,9	76	6,6	37,6	33,9
	А	131	94,6	49,4	77	6,4	37,3	34,5
	Б	131	99,3	48,0	79	6,5	36,8	34,4
2011	И	132	93,2	46,7	73	6,5	36,6	34,0
	А	131	97,8	45,1	75	6,6	37,4	34,6
	Б	131	98,3	47,0	72	6,6	36,1	34,3

Table 23

Fiber yield by families of comparable elite material selection methodologies,%

Year experiences	Variant techniques	N	X	&	V	&x	&x%
2009	И	101	34,2	1,6	4,6	0,1	0,3
	А	73	34,3	2,6	7,6	0,3	0,9
	Б	81	34,5	2,1	6,0	0,2	0,6
2010	И	181	35,7	2,4	6,7	0,2	0,6
	А	52	36,0	2,8	7,6	0,4	1,1
	Б	61	35,6	2,3	6,5	0,4	1,1
2011	И	178	32,7	1,9	5,8	0,1	0,3
	А	90	34,1	1,7	5,0	0,2	0,6
	Б	92	33,9	1,9	5,6	0,2	0,6

Table 24

Fiber length by family of comparison elite material selection techniques, mm

Year	Variant	N	X	&	V	&x	&x%

of experience	techniques						
2009	И	101	34,0	1,6	4,7	0,2	0,5
	А	73	34,0	1,2	3,5	0,1	0,2
	Б	81	34,2	1,2	3,6	0,1	0,3
2010	И	181	32,3	1,4	4,3	0,1	0,3
	А	52	32,3	1,4	4,3	0,2	0,6
	Б	61	32,5	1,4	4,3	0,2	0,6
2011	И	178	33,0	1,4	4,2	0,1	0,3
	А	90	33,2	1,3	3,9	0,1	0,3
	Б	92	33,2	1,2	3,9	0,1	0,3

§4.6 Development of a methodology for the production of original cotton seeds of new and registered cotton varieties

Under the existing methodology of elite seed production, the authors of the variety are practically not involved in the selection of original seeds. This leads to the fact that in 5-6 years populations of the variety appear. The proposed method will allow to preserve the original properties of the variety for a long time, reduce the cost of production of elite and in a short period of time sharply increase the seed yield. In addition, the new method facilitates the work of the author to improve the variety, which may contribute to the extension of its cultivation in production. When carrying out purges, it is necessary to make fuller use of modification deviations in productivity, but at the same time it is necessary to rigidly reject hereditary changes, removing underdeveloped, diseased and reduced forms. Differential use of different methods, taking into account the biology of the variety, rational simplification of the scheme of production of seeds of super-elite and elite in our opinion will save the type of variety, accelerate the production of elite, reduce its cost in 1.5-2 times, increase the seed multiplication factor. The produced seeds should not be distributed, but sold with a guarantee of their quality. The proposed method will allow to preserve the original properties of the variety for a long time, reduce the cost of production of elite and in a short period of time sharply increase the yield of high quality seeds.

Table 25

Results of 1-year nursery field screenings using the existing methodology

Reason for	1 Field View			2-3 Field			Total rejected							
	Total number **of**			Total number **of**			**Semey**		Plants in **them**		Plants in non-		Total **plants**	
	number	**percentage**	Growing up in non-domestic	number	**percentage**	Growing up in non-domestic	**quantity**	**percentage**	**quantity**	**percentage**	**quantity**	**percentage**	**quantity**	**percentage**
By **atypicality**	**231**	23.1	**751**	**5**	0.5	.	**236**	23.6	**7400**	21.5	**751**	2.2	**8151**	23.6
Including **sterility**	-	-	**28**	-	-	-	-	-	-	-	**28**	0.08	**28**	0.08
By defeatability:	-	-	-	-	-	-	-	-	-	-	-	-	-	-
Wiltom	-	-	**11**	-	-	- 7	-	-	'	-	**18**	0.05	**18**	0.05
nests	-	-	**1608**	-	-	-	-	-	-	-	**1608**	4.7	**1608**	4.7
Thinning	**29**	2.9	-	-	-	-	**29**	2.9	**478**	1.4	-	-	**478**	1.4
On **drying**	-	-	-	**21**	2.1	-	**21**	2.1	**632**	1.8	-	-	**632**	1.8
Late maturity	-	-	-	**78**	7.8	-	**78**	7.8	**1909**	5.5	-	-	**1909**	5.5
By low **yield**	-	-	-	**55**	5.5	-	**55**	5.5	**1413**	4.1	-	-	**1413**	4.1
ALSO:	**260**	26,0	**2398**	**159**	15,9	7	**419**	41,9	**11832**	34,3	**2405**	7,0	**14237**	41,6

Table 26

The results of the nursery's 2 year old field viewings
by current methodology

Reason for **rejection**	1 Field **View**			2-3 Field **Viewing**			Total **rejected**							
	Total number **of** rejects			Total number **of** rejects			**Semey**		Plants in **them**		Plants in non-		Total **plants**	
	families	**percentage**	Growing up in non-domestic	families	**percentage**	Growing up in non-domestic families	**quantity**	**percentage**	**quantity**	**percentage**	**quantity**	**percentage**	**quantity**	**percentage**
By **atypicality**	**101**	22,4	**928**	3	0,7	-	**104**	23,1	**18552**	22,1	**928**	1,1	**19480**	23,3
Including **sterility**	-	-	**43**	-	-	-	-	-	-	-	**43**	0,05	**43**	0,05
By defeatability:	-	-	-	-	-	-	-	-	--	-	-	-	-	-
Wiltom	-	-	**10**	-	-	2,6	-	-	-	-	**36**	0,04	**36**	0,04-
pests	-	-	**4666**	-	-	-	-	-	-	-	**4666**	5,6	**4666**	5,6
Thinning	**20**	4,4	-	-	-	-	**20**	4,4	**2457**	2,9	-	-	**2457**	2,9
On **drying**	-	-	-	**18**	4,0	-	**18**	4,0	**3067**	3,7	-	-	**3067**	3,7
Late maturity	-	-	-	**23**	5,1	-	**23**	5,1	**3786**	4,5	-	-	**3786**	4,5
By low **yield**	-	-	-	**17**	3,8	-	**17**	3,8	**2883**	3,4	-	-	**2883**	3,4
ALSO:	**1217**	26,9	**5647**	**61**	13,6	**26**	**182**	40,4	**30745**	36,7	**5673**	6,8	**36418**	43,5

According to the proposed method, sowing costs will be reduced Z times, laboratory evaluation is carried out once in 5 years, it is not necessary to prepare annual bagging. In the original nursery by the proposed method 6 purifications were carried out, in the seed nursery 4. The results are given in the tables.

Table 27

Data on sweeps in the original nursery using the proposed methodology

Reason for rejection	Date of cleaning						Total plants rejected	
	15.06	6.07	25.07	15.08	15.09	7.10	piece	%
By atypicality	-	5	12	161	74	17	269	3,9
Sterility	-	-	-	2	8	3	13	0,2
Defeatability: wilt	-	1	3	9	14	46	73	1,1
Pests	15	210	7	11	-	-	243	3,5
TOTAL:	15	216	22	183	96	66	598	8,7

Table 28

Seed nursery cleaning data using the proposed methodology

Reason for rejection	Date of cleaning				Total plants rejected	
	16.06	27.07	26.09	8.10	piece	%
By atypicality	-	9	198	73	280	2,8
Sterility	-	-	5	11	16	0,1
Lesions: wilt	-	2	42	57	101	1,0
Pests	42	698	4	-	744	7,5
TOTAL:	42	709	249	141	1141	11,4

These studies lead to the conclusion that under the conditions of market economy, only through the introduction of new methodology, without increasing production costs, there will be an increase in the seed multiplication coefficient, which will contribute to reducing the number of seed multiplication, will lead to a reduction in the number of elite seed farms, reducing the cost of preparation of super-elite and elite seeds. And most importantly, the author- originator of the variety can control the movement of seeds from the very beginning of reproduction, which will allow him to stipulate his requirements for remuneration for the use of breeding achievement when drawing up a license agreement with the seed producer.

The remuneration can be paid only if the seeds meet the requirements of standards. Based on the conducted research we have developed a methodology for the production of original seeds of new and registered cotton varieties.

§4.7. Study of genetic and phenotypic homogeneity on varietal alignment

The cotton industry of the Republic is actively developing and is one of the key links of the economy, taking the sixth place in the world in the production of cotton fiber. Therefore, systematic study of the global and domestic cotton market conditions, development of proposals for rational placement of cotton varieties is very relevant.

The efficiency of varietal renewal in the placement of varieties mainly depends on the level of elite seed production, i.e. on the organization of primary seed production, which includes the first links preceding the cultivation (multiplication) of super-elite seeds, which includes the selection of source material, its evaluation and multiplication. The quality of cotton products and yield potential of super-elite, elite and subsequent reproductions are equivalent and do not decrease in the process of reproduction under the 5-year varietal renewal scheme.

Deterioration of a variety is the process of reduction of its economic and biological qualities on the basis of hereditary variability due to splitting, appearance of mutations, mechanical and biological contamination and reduction of resistance to seed-borne diseases.

The cultivation of varieties under low agronomic practices is often cited as one of the reasons for variety degeneration. But we cannot agree with this. Low agrotechnique itself, as well as any other external influence, cannot cause adequate deterioration of heredity. But at low agrophos, unfavorable conditions are created for the formation of sowing and physical qualities of seeds, and when sowing with such seeds, the hereditary qualities of the variety cannot be fully realized.

Obtaining pure seed is a problem that involves not only varietal traits, breeding and seed production methods, and the influence of various development conditions, but also the precautions that are necessary during harvesting, ginning, and seed transfer. Therefore, rational placement of varieties contributes to meeting the demand for cotton fiber in the world market of the required quality and allows organizing seed production work at a high level.

Cotton variety in the process of reproduction should have stable indices of economically valuable traits and preserve the homogeneity of morphological traits.

In elite-seed crops of released cotton varieties, along with obvious biological and mechanical impurities, plants deviating from the variety description only by individual morphological traits are found. Phenotypic difference of plants depends on agro-technique, plant growing zone, year conditions and genetic background created by biotypes of the variety. Depending on which biotypes predominate in a variety, elite materials within a nursery may differ. By rejecting atypical plants in nurseries, complete homogeneity of morphological traits of plants is not achieved; each year typical families and plants have to be selected and atypical ones rejected. In this case, not all genetic variability of the population is used, but only some part of genotypes, and the initial heterogeneity of the population is changed, its competitive ability is reduced, and the efficiency of selection of elite material in nurseries is reduced. On the other hand, biological contamination of a variety leads to its degeneration. Since 2009, we have started to study the comparative evaluation of the efficiency of individual selection in different released cotton varieties on the example of varieties An-Bayaut 2 and C-6524.

In order to determine the genetic diversity of elite materials, genetic dispersions, coefficients of genetic and phenotypic variability, coefficient of inheritance of traits were determined. The experiment was laid for the first time since 2009 as a source material for the experiment were allocated average samples of seeds from nurseries of variety C-6524 in Chinaz elite farm and variety An-Bayaut 2 in Gulistan elite farm. The test of the reporting year was conducted in sixfold repetition at 100 m plot and randomization of experiment variants in repetitions. In all repetitions the description of morphological features of each plant was made, the number of deviated forms was determined according to the features of the above-mentioned varieties. Individual selections were collected from the plants of each phenotype typical to the variety and deviated by morphological traits to check them by progeny. The data obtained for each trait in each nursery separately was processed by analysis of variance. After proving the significance of variant differences, the following genetic and statistical parameters were determined through the F criterion:

1. The coefficient of genetic variability:

V g = ------- 100%

where: g - genetic variance of /variance/

x - arithmetic mean for the given one

to recognize:

Genetic variance / 8 g/ was determined by the formula:

Where: - dispersion of families

- random variance

r - number of repetitions

2. Phenotypic variability coefficient

%= -------- 100%

3. The heritability coefficient /H / was determined by the ratio of population genetic variance / / to total phenotypic variance /H /.

A total of eight phenotypes were identified, seven of which we

categorized as
to the rejected ones:

1. An-Bayout 2 is a typical bush. The stem and fruiting branches are pubescent, while in variety C-6524 the stem and fruiting branches are slightly pubescent; in both varieties they are covered with anthocyanin tan in the fall. An-Bayaut 2 has 2 fruiting branches of the semi-second branching type and C-6524 of the semi-second branching type:

An-Bayout 2 variety has a rounded ovoid boll with a smooth surface, while variety C-6524 has an ovoid elongated shape.
The leaves of both varieties are medium-sized 3-5 lobed.

2. Spreading plants
3. Slender stem, elongated box
4. Fruit branches of the first type, boll is rounded
5. Compact bush
6. The box is light and rounded
7. The box is large
8. Thin branches, box rounded. Sterile plants, on general grounds, were classified as non-tropical.

In each nursery, 1081 to 1158 plants were analyzed, with typical and deviant forms identified. In each repetition of the experiment, 10 plants of each phenotype were selected from typical and deviated plants.

The harvest from the selected plants was collected separately.

Distribution of nursery plants by phenotypes is given in Table. As can be seen from the above data - in both tested cotton varieties the largest number of plants typical for all morphological traits (the smallest number of deviant phenotypes) was in the populations isolated in the seed nursery of the second year, i.e. in the seed multiplication crop.

The highest number of rejected phenotypes, also in both cotton varieties, was in the population selected in the seed nursery of the first year, corresponding in the field to the elite material of the nursery. This confirms the theoretical position of increasing varietal purity of elite variety materials at the final stage of elite production and shows the unacceptability of harvesting individual selections in the nursery of the second year, where in fact individual selections are harvested in elite farms.

As can be seen from the data, the coefficient of genetic and phenotypic variability of both traits in An-Bayaut 2 variety is several times, and the coefficient of genetic variability of boll weight is many times greater than in variety C-6524. Consequently, despite the close phenotypic homogeneity of the tested varieties in morphological traits, they differ significantly. In terms of variability of economically valuable traits, in particular, raw cotton weight and boll weight of individual selections.

Greater genotypic variability of economically valuable traits of plants of An-Bayaut 2 variety caused greater theoretical heritability /H / of both traits in comparison with variety C-6524. The variance analysis of data on total yield is shown in Table 3. 3, which shows that in variety C-6524 the differences in trait variability between nurseries were less significant, while in variety An-Bayaut 2 the criteria were actually below the theoretical significance level of 0.010 and 0.032. Thus these data indicate greater homogeneity of plants typical to variety C-6524 in terms of productivity of An-Bayaut 2.

The highest coefficients of variability of An-Bayaut 2 variety in the nursery of the first year and low in the nursery of reproduction. Due to less homogeneity of plants of An-Bayaut 2 variety in productivity, variability of this trait from year to year (in nurseries) is sharply reduced, while variability of productivity of typical plants in nurseries is much more stable in variety C-6524.

For the nursery of the first year, An-Bayaut 2 variety, the coefficient of genetic variability of typical plants / =1363.2/. It is higher than in deviated plants / = 889.1/. By nursery of the second year, on the contrary, deviated plants / = 1043,2 / more than typical plants / = 812,1 /. By nursery breeding typical plants / = 635.1 / more than deviated plants / = 165.7 /. The difference between the coefficients of phenotypic variability. Typical / = 46.4 / and deviated / = 44.9 / plants of the nursery of second year family trials deviated plants / = 50.6 / is greater than typical / = 36.1 /. In the breeding nursery of typical plants, / = 30.6 / more than that of outliers / = 21.4 /.

It is proved that phenotypic coefficients of variability do not always reflect the genetic homogeneity of varieties, which reduces the reliability of assessment of genetically determined differences between elite materials and the efficiency of selection. Heritability coefficients H of typical plants in the second and first year family trial nurseries and in the propagation nursery were 0.05; 0.044; 0.54, respectively, whereas those of outlier H were 0.030; 0.010; 0.032, respectively.

In the nursery of the first year family trials, the coefficient of genetic variability of typical plants / = 784.1 / is greater than that of deviated plants / = 724.1 / in variety C-6524. In the nursery of the second year family trials, the opposite is true; deviated plants have / = 722.8 / more than typical plants / = 490.7 /.

According to the nursery trial of first year families of variety C-6524 the coefficient of phenotypic variability of typical plants / = 48.9 / more than in deviated plants / = 38.7 /. And in pythium-
The number of second year family trials in deviated plants / = 41.6 / is greater than in typical plants / = 33.8 /. In the breeding nursery of deviated plants / = 43.3 / more than that of typical plants / = 38.1 /.

The data obtained show that there are large differences between the coefficients of genetic and phenotypic variability. The latter do not reflect the really existing differences between varieties in terms of their degree of equalization. According to the theoretical inheritability of traits / H / regular data in the experiment of the reporting year were not obtained, but it should be noted that at high coefficients of variability inheritability of yield is very low. Low inheritability of productivity was found earlier by other researchers. This is reported by foreign researchers B.Christiris, J-Harrison /1959 /, N.G.Simongulyan /1970,

1974, 1975 / and others. Low inheritability of productivity is explained by high paratypic variability of this trait. At the same time, attention is drawn to the regular growth in the years of reproduction of elite average population / x / productivity of typical plants in both varieties, especially C-6524. At the same time, in deviated plants of both varieties, the average population productivity of plants also decreased naturally over the years of elite reproduction. The main requirements that varieties should meet are a high degree of adaptation to the conditions of their intended growing zone, specified parameters of productivity, quality, resistance to abiotic and biotic stresses, stability of yields under unstable hydrometric conditions.

CHAPTER V. INBREEDING METHOD FOR SEED PRODUCTION (INDIVIDUAL SELECTION COMBINED WITH SELF-POLLINATION) IN PRE-BREEDING

§5.1 Effect of self-pollination on varietal purity and longevity of the variety

It is known that seeds are carriers of biological and economic traits of plants, their quality largely determines the yield of agricultural crops. In the process of reproduction, due to intensive natural selection, a variety becomes more and more adapted to local growing conditions, which in many cases increases its vitality. In addition, favorable growing conditions (agroclimatic complex) can cause long-term modifications, which, overlapping from year to year, also create prerequisites for improving the biological characteristics of the variety.

Preservation of these useful qualities of a variety in the process of its multiplication and production use is the main content of seed production work. To solve this problem, it is necessary to develop new and further improve existing methods and techniques of seed production. This, in turn, necessitates deeper and deeper study of the nature of those biological phenomena, the skillful use of which predetermines the successful solution of modern seed production problems.

In cotton, unlike other agricultural crops, the basis of varietal renewal is elite farming. The main objectives of elite-seed farms are: preservation of typicality, the best economic-valuable traits and technological properties of fiber, production of elite and first reproduction seeds with high varietal and sowing qualities necessary for varietal renewal. The current Instruction on production of elite and first reproduction seeds of the cotton varieties of zoned cotton does not provide for elite-seed breeding work taking into account the genetic features of varieties, the outcome of their breeding and selection, the duration of being in production.

Based on the concept of degeneration of self-pollinated varieties, it is suggested to increase the volume of sampling lines to impoverish the heredity of varieties, to conduct intensive scraping and long-term testing in order to exclude the "worst" offspring in the elite. In the process of using even a well-selected variety, the economic and biological traits characteristic of the variety gradually decrease and it deteriorates. This is due to mechanical and biological contamination, splitting and increase of seed-borne diseases.

The seed production process does not fully comply with the principles of genetics, i.e. progeny testing, heredity and variability of a variety. The basic principles of work in primary seed production, the scale of selection of elite plants for variety reproduction, methods of evaluation and intensity of line rejection, special methods of maintaining its valuable properties, as well as the features of variety reproduction in self-pollinated and cross-pollinated crops have not been fully developed.

How does the life of a variety develop after release, how long does it last, and what factors determine the life of a variety and whether it can be extended? We consider these questions on the example of the organization of cotton variety renewal in elite-seed farms.

Elite-seed work requires a daily creative approach. The elitist, like a breeder, systematically reviews the crops, selecting the most typical plants and families based on the author's description, taking into account the conditions of the season and cultivation. When selecting seed material, knowledge of genetics - the science of plant heredity and variability - is needed, and a good knowledge of statistics is needed when comparing variation series.

Many managers of elite farms are well versed in the entire cycle of reproduction of elite seeds: from individual selection to seed multiplication offspring. But, unfortunately, a significant part of specialists of elite seed farms have poor knowledge of the methodology of elite work, make significant mistakes in rejection and selection of elite material.

Breeding institutions often conduct breeding work very effectively and intensively, but because there is no sufficient base, poorly carry out the development and multiplication of varieties and lines. The transfer to elite farms of obviously underdeveloped varieties and even biological mixtures is the reason for low varietal purity. In some cases seed breeders have to deal with erroneous author's description of varieties, which excludes proper selection of elite materials and preservation of variety traits. Multiplication of new varieties in seed farms is carried out without any spatial isolation. At the same time, cotton also has a tendency to crossing, which means that biological mixing is quite possible. Therefore, it is unlikely that under such conditions a seed farm will be able to refine a variety and bring its varietal purity up to the standard requirements. The offspring of hybrid plants give a large number of individuals of inferior quality, which leads to noticeable degradation. In this sense, the variety does degenerate rapidly, depending on the size of the mixing that takes place.

Proper seed production is based on the use of genetic regularities and knowledge of the biology of cultivated crops and varieties. A variety in its mass consists of plants that are homotypic in morphological traits and economic and biological properties. The uniformity of plants within a variety is created by selection and maintained by self-pollination in self-pollinating crops. A variety can be considered as a self-reproducing, relatively stable discrete biological system. The degree of plant uniformity is determined by the constancy of the method of plant pollination and the level of modification variability. Cross-pollination by other varieties and crops equally removes the stability (uniformity) of varieties of both cross-pollinated and self-pollinated crops.

In genetic terms, self-pollination leads to the identification of recessive traits that are latent in the plant. This can be schematized as follows. If we imagine conditionally that as a result of cross-pollination the plant has received some additional recessive genes, then during 3-4 years of self-pollination (depending on the degree of heterogeneity of the initial forms), these alien recessives should easily manifest themselves in the form of traits that are not peculiar to this type of plants. Further self-pollination leads to stability of genetic traits characteristic of a variety or form.

Vegetable breeders in North America prefer selection in pure lines to selection in the initial heterozygous population or in early generations of hybrids. It is noted that Johansen's theory of pure lines has played an outstanding role in the development of many theoretical statements of genetics, but its application in breeding is often overestimated. European breeders pay less attention to the pedigree method than American breeders, although the latter have also begun to realize that a "pure line" (according to Johansen), fully homozygous, cannot be obtained even as a result of long inbreeding. The efficiency of self-pollination was pointed out by N.I.Vavilov, he directly oriented to the use of self-pollination in breeding for self-pollinators prone to crossing. Industrial cotton varieties in terms of productivity do not experience inbreeding depression even with very deep forced pollination. Many lines are not only not inferior, but also surpass the elite in productivity, a trait. Which has not only economic, but also a great biological and evolutionary value. In experiments under selection aimed at high viability and productivity, there was no noticeable depressive effect of long-term self-pollination in the obtained inbred lines of cotton. By means of long-term self-pollination and selection not only for homozygosity, but also for high viability and productivity, they obtained lines in which the depressive effect of long-term self-pollination was absent. The latter can be explained, apparently, by the fact that the degree of self-compatibility increased in these lines under the influence of selection.

In different cotton-growing countries, cotton seed production utilizes a number of methods of seed multiplication and seed supply. One of them is that before seed transfer by farmers (USA, ARE), plants are self-pollinated for a number of years: first, in the breeding nursery -3-5 years; then in the 1st year of multiplication on a plot of 0.2 ha, the progeny of the best 30-50 self-pollinated plants are sown, the seeds of which in the 2nd year are used for sowing on an area of 10 ha, where all measures are also taken to preserve the genetic purity of the material (protective rows of cotton or corn crops, corresponding to the location of the plots, etc.); in the 3rd year, multiplied seeds can be sown with 250,000 seedlings of cotton or corn, corresponding to the location of the plots, etc.).etc.); in the 3rd year, 250-280 ha can be sown with multiplied seeds; in the 4th year, the planned production areas. These stages are under the strict control of the breeder-author of the variety; although further multiplication can be entrusted to farmers, vigorous measures are taken in subsequent years to maintain the purity of the variety. In particular, the sowing of a single variety is recommended to prevent cross-pollination. Mono-variety also eliminates mechanical contamination of the seeds.

The main task in seed multiplication is to maintain the economically valuable traits determined by the author during the zoning of the variety. At the same time, special attention is paid to the preservation of varietal conditions of the variety, i.e. the produced seeds should meet the requirements of the standard in terms of varietal purity. Based on this, we decided to find out the reason for such a discrepancy and analyze the state of varietal purity of multiplied cotton varieties.

§5.2 Causes of reduced grade purity

One of the main reasons for the decrease in varietal purity of varieties and their loss of cotton fiber quality is that in cotton breeding remote hybridization is used, involving wild and semi-wild forms of cotton. The remoteness of genomes violates the general recombination process and the balance of the genetic system, cleavage occurs at many loci, stabilization of polygenic traits comes in very late generations. Therefore, varieties and especially those of polyploid nature based on distant hybridization should be refined for a long time and not introduced into production until the homogeneity of economically valuable traits reaches a certain limit. Otherwise, the loss of valuable qualities of the variety is inevitable.

One more aspect of this problem should be noted. Cotton breeding should be mainly based on forced self-pollination, which makes it possible to obtain constant homozygous forms in a short period of time and to keep them pure in the future.

Self-pollination helps to identify recessive alleles hidden in heterozygotes, as a result of which new forms homozygous for many traits can be isolated. Considering that the majority of economically useful traits are controlled by recessive genes, this method is naturally of great value.

In his experiments, J. Schell (1922) showed that forced self-pollination allows to get rid of many lethal genes, reduce variability of traits (coefficient of variation) and stabilize it at the same level by the 5th-6th generation. In the USA, breeding work with cotton is almost entirely based on forced self-pollination. Similar work is carried out at the first stages of seed production, and American varieties, as a rule, have a high level of morphological and genetic homogeneity. Therefore, it can be understood how important is the degree of genetic refinement of a new variety submitted to the State Variety Network and for multiplication.

The breeder, releasing unequalized elite, condemns it to degeneration in advance. Many specialists engaged in cotton seed production discuss the effectiveness of intra-varietal crossing. They also suggest that intra-varietal crossing in cotton seed production does not have a positive effect and suggest limiting it to directed selection with subsequent check on progeny.

Studies conducted in different years, in different zones and on different varieties show the effectiveness of intra-varietal crossing of cotton. However, the percentage of biological contamination of the variety of the subsequent generation in intra-sort crosses has not been studied so far, and there is still no complete clarity in assessing the biological significance of self-pollination.

Some breeders and seed breeders believe that the main reason for the deterioration of self-pollinating varieties in production is their prolonged self-pollination, so with each reproduction they seem to progressively lose their yielding qualities. But, based on other works, it is clear that in self-pollinating crops no degeneration depression or deterioration under the influence of self-pollination occurs.

Some scientists believe that a variety begins its life in production with the most typical elite plants. Being well-selected, it persistently retains its hereditary qualities over a number of generations.

In the process of elite reproduction, the economic and biological traits of a variety are constantly changing. This process is associated with splitting, appearance of mutations, mechanical and biological contamination, increase in plant diseases, etc. Due to the above reasons, in the long-term cultivation of subsequent reproduction, deterioration of varieties occurs in production. For more accelerated stabilization of economically valuable traits of the line and increase its homogeneity, we adopted a method of artificial self-pollination (inbreeding) of flowers on typical early maturing, productive with high quality and white fiber individuals with subsequent verification of offspring, rejection of undesirable plants and selection in families of forms. In the first year (2009), seeds from self-pollinated bolls (i1) progeny of early self-pollinated bolls (i 2) were sown. In the i 1 nursery, crosses between families were carried out to obtain interseeded F 1 hybrids (i 1 x i 1) in order to create a promising varietal population from the best combinations later on. This is essentially interbreeding, i.e. artificial pollination within the hybrid population. Academician I.S.Varuntsyan wrote about it back in 1970: "Instead of inbreeding, Andrus strongly advises to widely use interbreeding, cross-pollination within the variety, within the hybrid population to apply sibcrosses (artificial crossing between related plants of the same origin). The same opinion was held by D.V.Ter-Avanesyan (1973): "Creation of multilineage varieties-populations with balanced genotype is one of the promising methods of breeding". Even earlier B.Christidis and J.Harrison wrote about it : In order to avoid further undesirable consequence of pure line, it is possible to resort to artificial in-line crossing instead of self-pollination". Intra-linear crosses were studied by S.S.Sadykov and H.Ashurbekov (1976), who concluded: "inter-linear hybrids of genotypically different but phenotypically identical lines have higher viability and yield".

For more accelerated stabilization of economically valuable traits of the line and variety and increasing their homogeneity, we adopted the method of artificial self-pollination of flowers on typical early-ripening, productive with high quality individuals with subsequent verification of offspring, rejection of undesirable plants and selection of forms in families for self-pollination.

§5.3 Conducting self-pollination on fine fiber cotton

In our research, interfamily hybrids F 1 were compared with each other in terms of maturity, boll maturation rate, typicality, fiber color and quality, and holoseminity. The best combinations were selected among them, which later formed the basis of the nucleus of the future breeding variety. At the same time, the progeny of plants obtained by artificial self-pollination (i 1 - i 3) was studied. In this case, seeds of the best individual selections were taken for subsequent sowing. Thus, seed development of families of inbred plant progeny was carried out by the classical method - selection of individual plants with verification by progeny. Consequently, we used genetic methods - inbreeding and interbreeding - when forming a new variety. Thus, we sought to prevent cross-pollination, accelerate the stabilization of economically valuable traits, and reduce the duration of the breeding process. It is known that in the USA cotton breeding work is based on forced self-pollination (inbreeding), which makes American varieties, as a rule, morphologically and genetically homogeneous.

In the first nursery, seeds were sown from boxes on which inbreeding, i.e. the progeny of inbred plants (i 1 -i 3), or systematic self-pollination, had been carried out in the previous year.

In the second nursery seeds were sown from self-pollinated boxes of individual selection (plants) and represented the offspring of one, two and three times self-pollination. The planting scheme was a single-row plot of 6 p/m, placing plants at 15 cm intervals with 60 cm row spacing.

The following parameters were taken into account: length of vegetation period, boll size, productivity, fiber yield, its color and quality according to the "Spinlab" system. The variety Termez-31 was sown as a standard, and 9871-I was sown for fiber quality.

In 2009, 2 nurseries were established: propagation and station variety trials. The total area of the trials was 0.57 ha. Sowing was carried out manually on May 14, or 20-25 days later due to weather anomalies of this season. According to the meteorological service, in April and 1-2 decades of May precipitation was more than 120 mm, i.e. in the most optimal time for sowing cotton there was a large amount of precipitation, which did not allow to carry out timely field work, including sowing of cotton.

During the period of cotton vegetation the following agrotechnical measures were carried out in the experiments: post-sowing loosening of furrows with rakes (breaking of soil crust); hoeing - 3, cultivation - 4, mineral fertilization - 2, vegetation irrigations - 4, chasing of the main cotton stalk on August 5. The mentioned agro-measures were carried out timely and qualitatively. In spite of this, the development of plants and, especially, opening of bolls was significantly delayed, especially in comparison with previous years. The reason for this was a significant delay in sowing dates and somewhat slow accumulation of the sum of effective temperatures during the whole vegetation period of cotton. Somewhat lower daily temperatures during the season significantly affected the slowdown of growth, development and maturation of cotton, especially in fine fiber.

However, early ripeness of breeding material, timely carrying out of chasing allowed opening of a part of fruit elements and by October 25 to complete harvesting works on the experiments. The results of research for each of the nurseries are given below.

The following materials were propagated in this nursery:

- on 68 rows - seeds of self-pollinated bolls (j1);

- on 204 rows - progeny of plants previously self-pollinated (j2);

- on 79 rows - individual selections from station variety trials ;

- on 95 rows - progeny of plants with cleistogamous flower type.

The total number of rows of the line in the nursery is -449.

Self-pollination of flowers was carried out on typical well-developed plants obtained from seeds of self-pollinated bolls . or in the above-mentioned 68 nursery rows (j2). In addition to self-pollination on these rows and plants, crosses between the best families of a given line were carried out to obtain the best varietal population. Hybridization was carried out on 28 combinations. This will select the best combinations. The selected varietal populations will become the basis, the nucleus of the future new breeding variety. During the harvesting period, raw cotton from bolls obtained by self-pollination were collected within each family (row). At the same time, hybrid (F0) bolls were collected for each hybrid combination separately.

The remaining rows of the line breeding nursery were harvested for raw cotton as follows:

Due to the significant lag in the development and maturation of cotton, the number of opened bolls was insignificant (5-6), which did not make it possible to collect individual selections, the harvest was conducted in a slightly different way than traditionally. Initially, plants were selected in each family where boll collection was possible. The best plants were selected in terms of development, typicality, set of fruit elements, boll size, not affected by macrosporiosis and wilt. Full bolls were collected from these plants, which amounted to 10 boll samples. After 2 weeks, the remaining opened complete bolls were collected from the same plants in one bag, i.e. individual-mass sampling. Thus, for the selected family we had 10 boxed samples and seed raw cotton of individual-mass collection. Thus, we saved the seed material in such an atypical, very unfavorable year for cotton. The volume of collected seed material on breeding line is: self-pollinated bolls - 47 bags, 10 box samples - 206, individual mass collection - 628.

Large-scale inbreeding of flowers of plants placed in 313 single-row plots, or families, was continued. Selected typical plants were marked with tags before self-pollination. After opening the bolls, (self-pollinated) raw cotton was collected in a row bag. A total of 304 bags were collected.

In order to maximize seed multiplication, seeds of individual selections and seed collections of the previous year were sown on 1063 rows. During the growing season, and especially from the moment of bolls ripening, the families and then the best plants were reviewed. Special attention was paid to the following traits: bush top pubescence, leaf color, boll shape and size, fiber color, fiber quality (length and fineness) and holoseminity.

Attention was paid to the productivity and maturation rate of bolls. If the previous morpho-economic traits are indicative of typicality, in addition to everything else, they are also of economic importance.

Holoseminity is the most important biological and economic trait. Raw cotton of holo-seeded varieties is easily divided into seeds and fiber during processing at cotton mills due to weaker attachment of fibers to the seed skin. Hence - less energy consumption per unit weight of fiber, higher labor productivity, less wear and tear of equipment (gin), which in general reduces labor costs and the cost of fiber produced. Besides, bare seeds exclude the necessity of additional expenses for their delintering, allow directly to carry out their chemical treatment, as well as to carry out sowing with precision seeders, which leads to reduction of seed sowing norms per 1 ha, and also reduces expenses for thinning of cotton seedlings. Finally, when processing raw cotton of holo-seed varieties, the fiber is less deformed and better retains its marketable appearance and natural fiber properties, which largely determines the consumer demand and affects the cost (pricing) of fiber.

In the experiment, due to uneven soil fertility, unevenness of the experimental plot, the development of plants was uneven. Hence, the plants showed some differences in the height of the main stem. Variation in productivity was also observed. During the growing season of cotton it was observed in some part of the plot neighboring with alfalfa crops that fruit elements in the middle part of the bush fell off due to alfalfa bollworm infestation. Despite these circumstances, a significant portion of the crops were in good condition, which allowed the best plants to be selected and raw cotton to be harvested. The number of individual selections harvested amounted to 2500 pieces. This fully ensured subsequent seed multiplication over a considerable area .

In the nursery inbreeding of the first generation progenies (j1) were placed in 99 rows. During the growing season, artificial self-pollination of flowers, also marked with tags, was carried out on the best plants. After bolls opened, raw cotton from these bolls was harvested separately. A total of 398 families, inbreeding progeny (j2-j4), were sown in the breeding nursery of individual selections. During the growing season, families were browsed, individual plants and the best families were evaluated. In this nursery, 650 individual selections of raw cotton were isolated and harvested.

Along with conducting self-pollination on fine-fiber cotton varieties and lines, simultaneous studies were conducted on medium-fiber cotton.

§5.4 Variability of economically valuable traits under repeated self-pollination

Inheritance of traits in cotton depends largely on genetic homogeneity and the degree of paratypic variability of traits. Cotton is a non-strict self-pollinator and is subject to natural crossing in the presence of insect pollinators and other certain conditions.

In pollination without castration of plant flowers, the percentage of cross fertilization reaches 40-80 depending on the variety. In the practice of seed production, under conditions of multisorting, biological contamination of varieties is especially possible. All these issues are of great importance for the methodology of cotton seed production.

In this regard, we set the task to study the degree of trait variability in cotton varieties under conditions of self-fertilization, cross-pollination and open flowering.

Elite seeds of the varieties included in the experiment were sown in 50-well plots and self-pollination and over-pollination were conducted for each variety within 10 days, from the beginning of flowering.

In the fall, all mature bolls from self-pollinated and cross-pollinated flowers were collected for each variety separately. The raw weight, number of seeds, fiber yield and length were determined for each boll.

For progeny studies, seeds from each box were sown separately as a line. For each variety, 30 self-pollinated and 30 re-pollinated lines were sown. Self-pollination was repeated for the same lines each year during 2008-2011. For each variety self-pollinated in previous years, forced self-pollination was again carried out by isolating buds on the eve of flowering with paper bags.

In order for the flower to develop normally, the insulator bag was made of thin, well translucent tissue paper. The size of the bag allowed the flower to open freely. In addition, to ensure that pollen germination conditions were the same for self-pollination and cross-pollination of the studied cotton varieties in the "intra-varietal cross-pollination" variant, the pollinated flowers were also covered with paper bags. The results showed that heterogeneity of varieties on qualitative dominant traits is revealed in the first two years of self-pollination. Heterogeneous plants were culled.

All self-pollinated lines of the studied cotton varieties mainly retained the typicality of the variety. However, compared to the overpollinated lines, the heterogeneity in some quantitative traits was more significant. It was revealed that the depressive effect of self-pollination affects first of all boll setting and boll weight reduction in the year of self-pollination (Table 29).

In all varieties, seed setting deteriorates equally depending on the variety. The decrease in the number of set seeds in the bolls of self-pollinated lines, as already mentioned, is the result of insufficient fertilization.

In self-pollination and over-pollination, incompatibility between pollen grains and pistil of self-pollinated lines plays an important role. Numerous studies show that the percentage of cotton seed setting is directly dependent on the number of germinating pollen grains and pollen tube growth. A similar dependence has been established by us.

Table 29

Seed setting of cotton varieties under self- and cross-pollination

Sort	Number of seeds set in the boll, pcs		
	By self-pollination M ± m	With overpollination M ± m	Deviations from self-pollination, %
Sultan	24,0±1,2	34,0±2,0	41,6
Zharkurgan	24,0±1,1	27,0±1,0	12,5
C-6524	23,0±2,0	26,0±1,3	13,0
Namangan-77	22,0±1,3	32,0±1,1	45,4

Pollinated and self-pollinated flower stalks 24 hours after self- and cross-pollination were fixed in 90% ethanol solution, and then the number of pollen tubes that passed through the stalk bases in the ovary was determined in cross sections stained with potassium iodide. During the study, the number of sterile columns and pollen tubes that did not pass through the base of the latter was counted.

Table 30

Pollen tube growth in self- and cross-pollenization pollination of cotton varieties

Sort	Number of columns, pcs.		Number of pollen tubes,
	total	sterile	

				passed into the ovary, pcs.(M m m)	
	S.o:p.o.	s.o.	p.o.	s. o.	p.o.
Sultan	44 42	11,4	no	42,0+3,9	149,0+4,7
Zharkurgan	39 34	20,5	no	29,0+4,1	109,0+7,5
C-6524	32 30	12,5	3,0	29,0+3,5	147,0+5,0
Namangan-77	37 30	10,8	no	32,0+3,4	100,0+5,6

Note. s.o., self-pollination; p.o., cross-pollination.

From the data of Table 33 it can be seen that the number of sterile columns increases markedly during self-pollination, in three varieties they were not found, and in variety C-6524 only three columns were sterile.

Along with high variability in such traits as the number of bolls on the plant, the number of seeds in the boll, the weight of the boll, self-pollinated lines differ among themselves in the degree of pubescence (pubescent, slightly pubescent), bush habitus (wider, relatively compact), but do not lose varietal typicality.

The cross-pollinated lines are homogeneous and show no deviation from the original variety.

The coefficients of variability in fiber yield and fiber length in self-pollinated and over-pollinated lines do not differ particularly, which indicates the relative stability of these traits. Their deterioration under the influence of self-pollination was not observed.

In 2010, varieties in the nursery of the second year and propagation were studied. Both variants were sown in parallel rows, and varieties were sown on the right and left: self-pollinated and non-self-pollinated. Plant productivity was determined by counting the number of bolls wilted on the bush on October 1.

It was found that there was no difference between self-pollinated and non-self-pollinated varieties. Both absolute values and coefficient of variability for the trait were identical.

In 2011, plants in the seed multiplication nursery were studied for yield, boll weight, fiber yield and length. The results confirmed the conclusion of 2010. However, noticeable deviations were observed in some varieties in boll weight, yield and fiber length. For example, in the variety Sultan, boll weight was higher in the non-self-pollinated variant - 7.8 g versus 7.2 in the self-pollinated variant. In variety C-6524, on the contrary, self-pollinated plants have a larger boll. However, there is no regularity: in one case, the self-pollinated variant is better, and in the other - the non-self-pollinated variant. A similar pattern is observed for fiber length, although in most cases plants of self-pollinated varieties and hybrids are better than non-self-pollinated ones. It is possible that these deviations are random in nature.

The current instruction on preliminary seed multiplication of new cotton varieties (Moscow, 1986) provides for taking test samples for technological fiber quality from the first and second bolls located on 2-4 fruit branches. At one time it was correct, then fiber strength and its grade (selected, first, etc.) were determined by them (samples). However, in connection with the transition of Uzbekistan to international standards of fiber quality, this indicator is not studied, and the concept of "micronaire" (Mic) is introduced, characterizing the fineness and maturity of fiber. The sample taken according to the current methodology significantly distorts the microneur index and significantly worsens the quality characteristic of fiber. In this case, the microneur index may be outside its basic range.

It seems that a more correct and more reliable methodology would be to take samples from the first harvest of raw cotton. As an example, here are the results of comparative analysis of fiber quality of the varieties of our station variety trials of the 2011 harvest (Table 31.).

Table 31

Results of comparative analysis of fiber quality of varieties of station varietal testing of 2011 harvest (Data of Uzbek Center for fiber quality certification "Sifat").

Sort	Variant	Fiber quality parameters				
		Mic	Len	Str	Rd	+b
St 9871-I	1	4,5	1,36	51,0	80,1	8,2
	2	3,8	1,27	38,4	69,9	11,0
Surhan-18	1	5,2	1,39	52,9	81,1	7,7
	2	3,7	1,37	38,0	70,0	11,2

Note: variants: 1 - the sample was taken according to the generally accepted methodology;

2 - the sample was taken according to the proposed methodology.

The table shows that the difference in fiber quality parameters between the variants is significant, especially in terms of micronaire parameters and specific breaking load. Indicators of the sample according to the attached methodology more reliably reflect the quality characteristic of fiber, which will certainly affect its pricing.

§5.5 Quality control of cotton fiber

Cotton fiber production is an industrial and agricultural industry . and cotton is one of the most important factors in the world economy. The main producers of cotton in the world are China, India, USA, Brazil, Pakistan, Australia and Uzbekistan, which together produce about 87% of the world's cotton. The largest consumers are China, India, Pakistan, Turkey, Bangladesh, where most of the world's textile production is concentrated. A high import flow of cotton goes to China, Turkey, Bangladesh, and Indonesia. Only two countries do not consume internally all the cotton produced - the USA and Uzbekistan. Together they account for up to half of the world's exports.

Cotton cultivation requires specific climatic conditions. The combination of high temperatures, dry air and abundant water is not common. Cotton is mostly grown in the deltas of large rivers. These are the deltas of the Mississippi, the great Chinese rivers, the Indus and Ganges, the Amu Darya and Syr Darya, and the Nile. In other places it occurs sporadically and does not play a major role in world production. With regard to the geography of cotton distribution in the world, it is necessary to mention the different varieties of cotton. Their names often reflect the area of distribution. Of the cultivated species, the Mexican species is the most widespread, accounting for about 70% of world production. Next comes the Indochinese species. Peruvian cotton gives the best fiber in length, fineness and strength. In Central Asia, mainly local varieties of Indochinese cotton are distributed, which are gradually being replaced by breeding varieties of Mexican and Peruvian cotton.

Depending on the location and growing conditions, different cotton varieties differ significantly in terms of fiber quality and properties. However, the variety of cotton is not the most important thing. The main thing is what kind of fiber it produces.

The limited area available for economically viable cotton cultivation means that there is little opportunity to increase the amount of cotton planted globally. The only resource for increasing cotton production in the world is to increase its yield. The only resource for increasing cotton production in the world is to increase cotton yields, as the area of land suitable for cotton is limited. In order to obtain high yields in industrial cotton growing, chemical preparations are actively used: pesticides (against various plant pests), herbicides (against weeds), fungicides (against fungal diseases), defoliants (used during harvest to shed cotton leaves). Cotton fields with conventional cotton are sprayed with approximately 20% of all pesticides and 25% of all insecticides in the world. All these substances are poisonous, resistant to decomposition and, once accumulated in the soil, cause irreparable damage to the ecology.

In order to reduce the cost of chemical reagents for pest control, research on genetically modified cotton was initiated in the United

States in the late 1990s. Cotton was one of the first crops to be developed. Genetically Modified (GM) seeds were used in its cultivation. Such seeds were first tried in Australia. Currently, about 26% of the world's cotton area is planted with GM or biotech cotton varieties, representing about 35% of total world production. The largest producer of genetically modified cotton is China. Australia, with which. In Australia, where the parade of GM cotton around the world began, all cotton is now genetically modified.

Annual crop rotation and avoidance of monocultures are the only reasonable alternatives for pest control. The cultivation of organic cotton, which is based on avoiding chemical fertilizers and using natural pollination, could be a truly sustainable success. Organic cotton cultivation eliminates the use of GM seeds. Currently, only a small amount of organic cotton yarn is produced in the world. The main supplier of organic raw materials is America.

Depending on the weather conditions during sowing and maturing of cotton and harvesting of raw cotton, as well as depending on the breeding variety, cotton fiber may have different micronaire parameters, staple length, color, blockage, strength (specific breaking load) and other quality parameters. According to these parameters, cotton fiber is divided into several different grades. In world practice, the seller and buyer in their large contracts, as a rule, do not limit the basic grade for calculating the value of cotton of different quality, supplied under this contract; in small contracts usually allow linkage to the official classification for the corresponding which do not make standards, are determined descriptively, based on physical standards.

One of the factors that had a great influence on the intensity of HVI technology development in the 1980s was the increasing use of rotor spinning machines in the world textile industry, for which fiber strength, which could not be assessed by conventional classing methods, was of paramount importance for the production of strong yarns. They were also used in USDA grading offices. At the same time, the system manufacturers Spinlab and Motion Control began shipping

HVIs to countries all over the world.

In Uzbekistan, the first HVI systems appeared in 1989-1990, and since 1993 the indicators were partially introduced into the national standard. Since 2002, the quality assessment of all cotton fiber produced in Uzbekistan is fully performed by HVI systems.

To perform measurements, the HVI system must be in standard climatic conditions of air temperature - 21 ± 1®C, relative humidity - 65 ± 2% (when monitored by Astman psychrometer with a scale of 0.1®C or equivalent temperature and humidity measuring instruments with accuracy between 6.75% and 8.25%.

All calculations are performed by the HVI's internal microprocessor software for each sample, with average measurement results for parallel test results. The final result of the cotton fiber measurements is output as a printer printout.

Thus, the analysis and assessment of the current situation in the sphere of production, consumption and quality control of cotton fiber has shown:

- significant changes have occurred in the production of cotton fiber;
- The main producers - China, India, USA, Brazil, Pakistan, Australia, Uzbekistan - currently produce conventional (natural) cotton using traditional agrochemical means (pesticides, herbicides, fungicides and defoliants) up to 60% of the total volume, genetically modified cotton (from GM seeds) resistant to pests and diseases - up to 35%, organic (environmentally friendly, without the use of chemical fertilizers) - up to 5%;
- The transition of cotton quality assessment in producing and consuming countries to international standards using the HVI high-performance measurement system has fundamentally changed fiber testing methods.

For more qualitative assessment of technological properties of fiber of individual selections, families of nursery I and II year in accordance with the conducted research, in our opinion, fiber sampling for analysis, to carry out the following way:

1)After cleaning individual selection, all the fiber obtained is sent for analysis;

2) for nurseries of I and II year fiber obtained after cleaning of seed collections is spread on the table in an even layer and 50 grams of fiber is taken from different places to be sent for analysis.

CHAPTER VI VARIETY CERTIFICATION (IDENTIFICATION) OF SEEDS

§6.1 Analysis of certification testing for the purpose of issuing a certificate of identification

Certification of seeds and planting material of agricultural plants is aimed at permanent control over production, harvesting, processing, storage, sale, transportation and use of certified seeds, as well as harmonization of the certification process with the rules and requirements of international organizations.

The object of certification is seeds of any species intended for sale and included in the State Register. The certificate is issued for seeds, which by variety and sowing qualities meet the requirements of state and industry standards.

In the process of reproduction, the economic and biological traits of a variety gradually change. This process is associated with splitting, appearance of mutations, mechanical and biological contamination, increase in plant morbidity, etc. Due to the above reasons, deterioration of varieties may occur during long-term cultivation in production.

Organizations for Economic Cooperation and Development (OECD) develop specific variety certification schemes for seeds in international trade. This allows the seed process to be monitored to ensure that the appropriate technical practices are followed to ensure the integrity of the variety. Checks are carried out at various stages of production to ensure that there are no unforeseen accidents that could reduce the varietal quality of the varieties being tested.

The introduction of variety certification schemes will allow to intensify seed trade, expand the areas of cultivation of crop varieties, enter the international market of domestic varieties and deepen international cooperation in the field of breeding and seed production.

Varietal certification of cotton seeds has not been carried out in our country so far. Instead, seed sowing approbation is carried out in order to provide crops with seeds of the best varietal and economic qualities. On the basis of approbation data, the varietal purity of the crops and the degree of pest and disease infestation are established.

The most important element of certification, as it is customary in international practice, is soil testing. In this case, it should be taken as a rule that if the results of tests on a batch of seed show that the variety traits or varietal purity are not preserved, the Certification Body has no right to certify such seed.

The Law of the Republic of Uzbekistan "On Seed Production" introduced certification of seeds on indicators certifying their varietal and sowing qualities. Seeds intended for sale within the country, as well as supplies to insurance funds are subject to certification.

Due to the fact that the Republic does not carry out variety certification and does not issue a certificate of identification, which is replaced by the act of approbation, we set the task to study the possibility of variety certification by inspecting fields and issuing a variety certificate for variety homogeneity.

§6.2 Research on varietal certification through seed crop inspection

Research work was conducted in field and laboratory conditions in Namangan region elite-seed farm "Chust" multiplying variety Namangan-77; in Fergana region elite-seed farm "Toshlok Super Elite" multiplying variety AN-16; in Yukari-Chirchik district of Tashkent region elite-seed farm "Guliston Darkhon Khosili" multiplying variety C-6524; in Akkurgan district elite-farming farm "Abduvali Temur" multiplying variety C-6541; in Gulistan district of Syrdarya province, elite farm "Ulugbek-4" multiplying variety An-Bayaut, in Pskent variety testing site and Republican station of primary seed production and seed science of agricultural crops. Breeding varieties developed by domestic breeders served as raw material. To establish the difference between one variety and another, as well as to assess the homogeneity and stability, the values of individual traits were used.

The varieties, with which the comparison was made in the process of testing, are included in the State Register of varieties recommended for sowing in the territory of the Republic of Uzbekistan.

The simplest criterion for establishing distinctiveness was the consistency of the distinguishing feature. The number of comparisons that is sufficient to draw a conclusion was determined.

A variety is considered homogeneous if its variability, depending on the method of selection and the presence of deviating forms due to random contamination by mutations or other causes, does not exceed the level that allows to make its description and identification accurately.

Homogeneity is the ability of a variety (breed) to exhibit its characteristic morphobiological features in almost all plants. The research is aimed at studying and harmonizing international legislation in the field of seed production. In the formation of yield properties and distinctiveness of cotton varieties of great importance are soil and climatic conditions of the regions of the republic, where they adapted to give a stable yield for many years.

The trials were conducted according to the methodology of the State Commission in fourfold repetition. As the results of the research showed, the majority of new medium-maturing cotton varieties being tested at the variety testing plots of the republic (in five regions) were significantly inferior to the more adapted to these conditions, zoned varieties of cotton. In addition, the trials in each region revealed several new medium-maturing cotton varieties that exceeded the standard varieties in terms of early maturity and yield. The individual new cotton varieties proved their superiority in these two traits in several regions.

The state of elite seed production can be judged by the final results, quantity and quality of sown seeds of elite and subsequent reproductions. Based on the volume of seed production, the state of affairs in primary cotton seed production could be assessed as satisfactory and even good. However, according to the data of ground control, many elite farms produce seeds with reduced varietal purity or non-varietal seeds instead of elite cotton varieties.

According to the latest reports of the Uzgoskontrolsemcenter, only a quarter of elite farms in Uzbekistan produced full-fledged elite seeds with 99% varietal purity. The remaining elite farms produced seeds with reduced varietal purity (at the level of 1, 11,111 reproductions) or even non-variety seeds (varietal purity 85-92%).

If non-variety seeds are released for a variety instead of elite, then all subsequent reproductions, i.e. all crops of that variety, become non-variety. Therefore, the release of inferior seeds by the majority of elite farms is the main problem of primary cotton seed production. In recent years, there are no varietal crops for some regionalized cotton varieties at all. There are many reasons for this situation. First, the reliability of evaluation and efficiency of selection of elite materials depend on the culture of farming in the farm. Low level of agrotechnics, different growing conditions within the nursery, overcrowding of elite-seed crops, which is often strived for by producers, untimely and poor quality of agro-measures during the growing season - all this makes it difficult to identify typical traits of the variety, and the transfer to elite farms of obviously underdeveloped varieties and even biological mixtures is the second reason for the reduction of varietal purity. In some cases, seed breeders have to deal with erroneous author's description of varieties, which excludes proper selection of elite materials and preservation of variety traits. The above examples indicate a certain underestimation of the role of varietal purity in the preservation of cotton variety traits and the unprofitable for the national economy of the country premature degeneration of released varieties. Since 1996, the determination of varietal purity of new varieties has been carried out at one of the state variety plots, first in Yukor-Chirchik, then in Pskent.

The implementation of this provision creates real prerequisites for establishing the work of elite farms on released varieties. Breeders are also interested in this, as objective evaluation of varieties allows selecting the best of them and reasonably placing them in cotton growing zones.

Regionalization and forced multiplication of improved varieties guarantees the preservation of their economically valuable traits.

According to recent reports, in many elite farms cleaning of elite seed crops from atypical plants is carried out untimely, poorly or not at all, and in some cases it is substituted by drawing up fictitious documents. Partially varietal conditions of seeds are reduced during cleaning of raw seed at cotton mills, but for high reproductions this contamination is usually insignificant. It is often overestimated with an understandable purpose - to shift the responsibility for the poor performance of elite farms to cotton mills. However, cotton plants are not and cannot be to blame for the low varietal purity of elite crops, since seeds for these crops are cleaned, as is known, not at cotton plants, but in elite farms.

The lack of control over varietal purity gives rise to the desire to increase the production of inferior seeds, to open additional elite farms. At the same time, seed production costs increase and seed quality deteriorates. One of the ways to improve seed quality is to carry out varietal inspection of seed crops and on the basis of its data to issue a certificate of indentification. This would be in line with international rules for seed certification.

6.3 Inspection of seed crops

Approbation of seed crops is carried out in order to provide crops with seeds with the best varietal and economic qualities. On the basis of approbation data the varietal purity of crops and the degree of pest and disease infestation of plants are established. Approbation is carried out within certain terms established by the Ministry of Agriculture and Water Resources, and the main document is the Approbation Act.

This scheme is not suitable for internationally traded seeds. The Organizations for Economic Co-operation and Development (OECD) have developed specific variety certification schemes for internationally traded seeds. They allow the monitoring of the seed production process to ensure that the appropriate technical practices are followed to guarantee the integrity of the variety. Inspections are carried out at various stages of production to make sure that there are no unforeseen accidents that could reduce the quality of the varieties being inspected. They include two essential ways to check that crop development is proceeding satisfactorily:

1. Seed samples are sown in field plots so that the plant can be monitored throughout the growing season.

2. Fields intended for seed production are inspected to determine their condition in some cases one or more times.

The purpose of field trials in control plots is to determine whether or not variety characteristics change during the breeding season.

The person responsible for making observations should have sufficient experience and knowledge of the variety and its characteristics to be able to make a qualified judgment as to the extent to which the plant differs from the sample and whether it can be considered atypical. Certain characteristics of the variety, such as height and maturity, are dependent on environmental conditions, causing differences between plants, so that there may be doubts about the inclusion of a particular plant in the count.

The test is designed to answer two questions:

-Whether the sample as a whole corresponds to the description of the variety;

- whether the sample conforms to approved grade purity standards.

To determine the authenticity of a variety by visual comparison and to identify atypical plants, plots should be arranged so that all samples of the same variety are placed together and within each variety all samples of the same generation, usually from the same batch of elite seed, are placed together. In this case, plots with different appearance and atypical plants are more visible and easier to observe. The second method involves establishing atypical plants in a plot to compare their numbers with approved standards.

Seed crops are inspected so that it can be ensured that there are no liabilities that could be detrimental to the quality of the seeds grown.

The main points to which the inspector should pay attention:

-is the sowing generally of the variety expected;

-whether the number of atypical plants exceeds the limits allowed by the standard;

- whether the number of plants of other species exceeds the limits allowed by the standard;

- whether the crop is adequately protected from mechanical mixing or extraneous pollination;

-Whether other aspects of plant health are satisfactory (absence of diseases and so on).

The inspector must give an independent opinion on the condition of the crop.

Inspection reveals the condition of the crop at the time of inspection, when some defects may be hidden or difficult to discern. In some cases a second inspection is required to make a final decision. But in all cases, the inspection is complemented by the results of tests on control plots, which are kept under observation and which usually provide more detailed information on the varietal and species purity of the seed stock.

The inspector should obtain information from the person in charge of the field about the location of the crop to be inspected, as well as any information: the amount of seed used for the crop, the yields of previous years in the field, and so on, which is normally recorded in the inspection report.

The person responsible for the seeding shall retain at least the label from the seed container used for seeding and present it to the inspector.

The inspector should be adequately trained to recognize the variety he is asked to inspect. He should be provided with appropriate descriptions of the main characteristics of the variety being inspected.

When evaluating a seed crop, a decision must be made as to whether or not it is in satisfactory condition and whether or not further detailed inspection should be continued. Seeds with significantly lodged plants, heavily weeded, stunted or weakly growing plants due to diseases, pests or other reasons that cannot be properly evaluated for varietal purity are discarded.

To simplify the work of inspectors we have developed draft model documents for different stages of field inspection. The first thing the inspector should familiarize himself with is the report of a legal or natural person on production of certified seeds, then the fields are inspected.

The purpose of the field inspection is to verify that the seed crop meets the standard for seed production:

-is that there are no circumstances detrimental to the quality of the seeds harvested;

-that the variety sown corresponds to the varietal traits;

- that the grade purity meets the requirements of the standard.

Seed crops should be inspected throughout the growing season. Formal inspection allows the evaluation of varietal authenticity and field cleanliness. Although field inspection techniques vary in detail, the main verification principles of field inspections are as follows:

1.Cultivated fields should be such that the risk of unwanted selfing plants or infestation of related seed crop species is minimized.

2.The crops should be sufficiently isolated from other varieties to prevent contamination by unwanted pollination.

3.Variety crops should be physically isolated from other varieties to prevent mechanical mixing at harvest.

4.Sown seed should be exclusively free of weeds and other varieties, especially those seeds that will be difficult to separate from other seeds during processing.

5.Sown seeds should be treated with toxic chemicals against diseases.

6.Sown seeds shall be of exact varietal authenticity, there shall be no presence of any other atypical plants in determining varietal purity standards.

7.There shall be no presence of plants of other species in the determination of standards.

The person conducting the field inspection shall be provided with all necessary information on the variety being sown and shall be familiarized with the traits used for varietal description and species identification. The necessary information shall include the variety description, the origin of the seed used for sowing and the results of the pre-inspection plots. Information on the cultivation of crops in the fields for the last five years should also be available to the inspector.

The inspector shall provide an independent opinion on the

inspected variety. The function of the inspector is to report to the authority on the condition of the crop during the inspection. If circumstances arise during the inspection in which it is very difficult to determine the presence of atypical plants or they become hidden, a second or subsequent inspection may be ordered until a decision is made.

For the purpose of authentication of seed sowing, seed producers must keep at least one label from the seed lot of the variety used for sowing. The seed certification body requires that one label be displayed in the field for the inspector to compare with a second label kept by the seed grower.

The purpose of this procedure is to check the data given on the label against that entered on the field inspection form and to declare the authenticity of the variety. After a complete field inspection, the inspector checks the fields in more detail, especially around the perimeter. The isolation of the crop around the perimeter must be in accordance with the norms. If the location of the variety, its authenticity, varietal identity, isolation and field treatment are all satisfactory, then the final stage of the inspection is the determination of varietal purity. In general, the number of seed sampling areas should increase in proportion to the size of the fields. Because of the high standards for super elite and elite seed, the number of plants to be examined should be larger than for reproductive seed. A sample size of 4n as a basic rule of thumb can be used when the required purity level is 1 in "n", such a sample for a minimum varietal purity of 99.9% (1 in 1000) the sample size should be 4000.

The certification body can mainly rely on the data from the pre-control plots and use the results of the field inspection only for confirmation. If there is a clear discrepancy between the control plots and the field data, a follow-up survey of both areas should be carried out so that a positive decision can be reached.

We have prepared some tables for inspection control, which are presented as numbers 29-32.

§6.4 Conducting inspection control

Inspection control was conducted in Chust, Yukari-Chirchik, Tashlak, Akkurgan and Gulistan elite-seed farms during 2012.

It was found that spatial isolation was maintained in all elite seed farms. However, atypical plants remained in the fields after field screening. Therefore, an additional varietal sweep with removal of atypical plants was recommended in all tested fields. After cleaning, it was allowed to collect seed stock.

In Y.Chirchik elite-seed farm of Tashkent region comparative analysis of rejection of plants and families during field screening by inspection control was carried out. For analysis 100 families in nursery of 1 year, 20 families in nursery of 2 years and 5 families in seed multiplication nursery were selected. Field screenings were conducted 3 times: during the periods of mass flowering onset, fruiting and before harvesting. Inspection control was conducted before sowing, after final thinning:June 1, July 1, August 25 (Tables 37-38).

The study showed that in the field screenings in 1 and 2 year nursery, the major percentage of rejection was due to rejected plants in rejected families. Out of 53.7% in the 1 year nursery and 43.4% in the 2 year nursery, 42.7% and 36.6% of plants were rejected in rejected families, respectively. This indicates that the elite farm workers pay little attention to quality field scraping for non-typicality. During inspection control, the percentage of rejected plants was 5-6 times lower; typical plants in thinned and poorly developed families were not rejected.

Table 32

COTTON TRAIT CHART

SIGN	EXPRESSION
1	2
1. CUST	
1 .Form.	Cone-shaped, Column-shaped, Narrow-pyramidal, Pyramidal, Wide-pyramidal, Globular

2.Compactness	Compact, Semi-compact, Sprawling
3 .Foliage	Strong, Medium, Weak
4.Number of monopods	Missing 1, 2, 3
П. STEBEL.	
5. Coloring	Light green, Green, Dark green, Red
6. Length of internodes	Short, Medium, Long
7 .Conformability	Resilient, Semi-resilient, Unresilient
B. Furrowing	Downy, Lightly downy, Naked
9.Degree of tan	Strong, Medium, Weak
Y. Cranking	Strong, Medium, Weak
W. SYMPODIAL BRANCHES	
1 1 .Type of branches	0 (marginal), 1, 2, 3, 4
12. Laying height of 1 sympodial branch	Low, Medium, High
13. Location of branches	Normal, Saggy
IV.SHEET.	
H. Magnitude	Small, Medium, Large
15. Swelling	Felted, Faint pubescence, Naked
16. Sheet surface	Strongly corrugated, Corrugated, Flat
17. Shape of the main lobe	Triangular, lanceolate, wedge-shaped.
18. Dissection	Strong, Weak, Medium
19. Shamrocks	Auricular, Lanceolate-lanceolate, Lanceolate.
V.COLOR.	
20. Magnitude	Large, Medium, Small
2 1 .Petal coloration	Light cream, Cream, Lemon.
22. Presence of anthocyanin stain	Yes, no.
VI.THE BOX	
23. Magnitude	Large, Medium, Small
24. Form	Cone-shaped, Oval-cone-shaped, Pyramidal, Ovoid-oval, Rounded-oval

25. Surface	Matte, Glossy, Smooth, Ribbed
26. Presence of a spout	No, Slightly pronounced, Strongly pronounced, Bent, Bent.
27. Presence of an asterisk	No, Mildly pronounced Strongly pronounced.
2 8. Sillage	3,4,5
	VII.CEMEHA
29.Magnitude	Large, Medium, Small
30. Swelling	Downy, plump, naked
31.Underbelly	No, Light gray with greenish tint, Gray, White, Emerald.

Table 33

Results of field screenings in 2014 for variety C-6524 in Yu-Chirchik elite farm

Reasons for rejection	Nursery 1 year 100 families				Nursery 2 years 20 families				Seed propagation nursery of 5 families			
	Semey	Raste total	Including.		Semey	Plants, total	Including.		Semey	Plants, total	Including.	
			In a bastardized family.	In unmarried families			In a bastardized family.	In unmarried families			In a bastardized family.	In unmarried families
Atypicality	20	442	400	42	1	404	371	93	-	97	-	97
Wilt Porosity:	2	94	74	20	2	703	660	43	-	43	-	43
Pests	-	130	-	130	-	119	-	119	-	50	-	50
Thinning	9	85	85	-	1	106	106	-	-	-	-	-
By underdevelopment	10	242	242	-	5	1402	1402	-	-	-	-	-
Bottom line:	41	993	801	192	9	2734	2539	195	-	190	-	190
%	41	34,1	27,5	6.6	4 5,0	47	43,6	3,4	-	1,8	-	1,8

Table 34

Result of inspection control in nurseries of Yu-Chirchik elite farm on variety C-6524 in 2014

Reason for rejection	Plant rejects											
	Nursery of 1 year, 100 families			Nursery 2 years, 20 families			Seed propagation nursery in families			Seeding of 1 reproduction 1 hectare		
	Quantity		%	Quantity		%	Quantity		%	Quantity		%
	Total number of plants pcs	Abandoned		Total number of plants pcs	Abandoned		Total number of plants pcs	Abandoned		Total number of plants pcs	Abandoned	
Atypicality		81	2,7		77	1,3		36	0,3		602	0,8
Sterility		-	-		-	-		-	-		-	-
Defeatability: wilt		34	1,2		173	2,9		67	0,6		920	1,2
Pests		117	3,9		201	3,3		108	1,0		2125	2,8

Bottom line:	**2988**	**232**	**7,8**	**6038**	**451**	**7,5**	**10950**	**211**	**1,9**	**75.6 t/ha**	**3647**	**4,8**

The following conclusions can be drawn from the data of the research on seed varietal quality evaluation norms:

1.International methods for determining the varietal qualities of cotton seeds need to be adapted to local conditions.

2.When describing a variety, the originator shall index the traits on the basis of which the plants are evaluated in determining varietal purity.

3.Tests on the distinguishability of one variety from another should be conducted in compliance with the agronomic recommendations recommended by the author of the variety.

4.In order to reliably evaluate the varietal qualities of the sown varieties, it is necessary to carry out seed crop inspection, and pay great importance to spatial isolation. A certificate of identification should be issued on the basis of the inspection and ground variety control data.

5.In order to establish the reliability of determining the varietal qualities of seed lots, it is necessary to plant control plots (ground control) and on them to check the homogeneity of the variety.

CONCLUSION

On the basis of comparative analysis of the accumulated scientific and practical experience in the development of elite seed production and adaptation of the most acceptable methods in the emerging conditions of seed production by farms of the republic:

- the process of statistical processing of results has been simplified

laboratory analysis of elite material;

- an optimal variant of reproduction of original and super-elite cotton seeds has been developed on the basis of methods of super-elite seed production applied in international practice;

- assessment of basic technological properties of fiber of trial samples and individual selections according to the international system HVI, generally accepted in the leading cotton-growing countries of the world, for these purposes the method of fiber sampling of individual selections, families of nurseries of 1 and 2 years has been developed;

- Improved accuracy and reliability of rejection results for selection of the best families of the propagated variety, which have quality indicators, according to international parameters, in order to maintain or improve fiber quality characteristics of the variety;

- development of a computer program for faster and more reliable evaluation of the author's material, for which a patent application has been filed, is in the final stage of formation and refinement of varieties;

- the optimal variant of modal selection of plants and families aimed at changing the variety structure in the desired direction, with strengthening of correlation in the corresponding group of plants was revealed.

For the first time on the basis of comparative study of the nature of inheritance, degree of variability and formation of economically valuable traits in different seed varieties, the genetic nature of the main economically valuable traits was revealed;

for the first time developed a computer software program "Elita", for statistical processing of the results of numerous laboratory analyses of elite seed breeding materials of cotton varieties and provided an automated process of rejection and selection of elite material for further multiplication;

justifications for reliable evaluation of individual selections and families of breeding varieties propagated in different elite farms are given;

in the practice of elite work it was proposed to use international norms of fiber quality assessment, and criteria for field rejection of families and plants were developed;

on the basis of comparative study of different methods of production of original seeds of new cotton varieties the instruction on preliminary multiplication of new cotton varieties was unified;

more reliable evaluation of plants and families in the nursery during field screening conducted on the basis of morphological table of traits given by the author when transferring the variety for variety testing was revealed and it was recommended to introduce it in elite-seed work;

It has been proved that under the current method of reproduction of new cotton varieties the authors of the variety practically do not participate in the selection of original seeds and often transfer for multiplication the source material with varietal purity corresponding to the quality of the 3rd reproduction;

when conducting field screenings, it is necessary to make more complete use of modification deviations in productivity, but at the same time it is necessary to rigidly reject hereditary changes, remove underdeveloped, diseased and reduced forms;

It is proposed to determine the fiber quality of individual selections and trial samples of seed collections of new breeding varieties in the pre-propagation farms on the territory of the Republic of Uzbekistan to analyze on the HVI line;

The advantage of harvesting individual selections in the nursery of seed multiplication of new cotton varieties has been established, which contributes to the preservation of genetic homogeneity and economically valuable features of reproduced seeds;

For the first time on the basis of comparative study of seed material evaluation by existing methods and with the use of HVI the spectrum of variability of economically valuable traits and technological properties of fiber in individual selections, nursery families of I. II year and seed multiplication;

matriolcal influence of the degree of harvesting of raw cotton seed on varietal and sowing qualities of seeds depending on boll collection to 8, 10, 12 sympodial branches was studied.

REFERENCE LIST

1.Encyclopedia of cotton production // Tashkent, 1985. -т.2. - C.127.

2.Law of the Republic of Uzbekistan. On seed production. - Tashkent: - 1996.

3.Law of the Republic of Uzbekistan. On breeding achievements. - Tashkent: - 1996.

4.Decree of the Cabinet of Ministers of the Republic of Uzbekistan: On the program of varietal renewal and varietal placement of cotton for 1999-2000 No. 491. -25.11.1998.

5.Resolution of the Cabinet of Ministers of the Republic of Uzbekistan: Main directions of seed production policy of the Government of the Republic of Uzbekistan. -№328. - 19.09.1996.

6.Regulations on the activities of state inspectors in the field of seed production of agricultural plants // J.: Breeding and Seed Production. -1998. - №4. -C.29-31.

7. Abzalov M.F. Evolutionary and breeding aspects of early maturity and adaptability of cotton in the studies of Academician S.S.Sadykov. // Mat.mezhd.nauchn.prak.konfer.posv.95 years. since the birth of Acad. S.S. Sadykov. - Tashkent: FAN.2005. C.12-13.

8. Abdullaev A.A., Saidaliev H., Khalmuradov A. Turlararo duragailashda guzaning chatishishish va chigit tugish kobiliyati // J.: Pakhtachilik. - Tashkent, 1996. C.7-9.

9. Abdullaev A.A. Historical aspects of the evolution of cotton early maturity. // Mat.mezhd.nauchn.prak.konfer. on the 95th anniversary of the birth of Acad. 95 years since the birth of Acad. S.S.Sadykov. - Tashkent: FAN.2005. C.5-9.

10. Abdullaev A.A., Klyat V.L., Rizaeva S.M. Evolutionary and historical aspects of natural and artificial selection for increasing the rate of cotton maturity. // Mat.mezhd.nauchn.prak.konfer.посв. 95 years. since the birth of Acad. S.S.Sadykov. - Tashkent: FAN.2005. C. 9-10.

11.Avtonomov A.A. Selection of thin-fiber cotton varieties. - Tashkent: Fan. UzSSR. 1973. C. 144.

12.Avtonomov A.I., Davshan A.S., Rodimtsev I.M., Shafrin A.N. Once again about intra-variety crossing of cotton // Cotton growing. 1965, №12. C.26-34.

13. Avtonomov V.A. Variability and inheritability of traits in hybrids with wild and ruderal forms of cotton. Avtoref.dis...k.S.kh. of sciences. - T.VNIISSKH,1983.-23 P.

14. Avtonomov V.A., Umbetaev I.A., Huseynov I.R. Genetic analysis of traits determining the early maturity of cotton species G.hirsutum L. // Mat.

International Scientific and Practical Conference "State of cotton breeding and seed production and prospects of development", dedicated to the 110th anniversary of Academician A.I.Avtonomov, 80th anniversary of Academician S.M.Mirahmedov and Professor A.A.Avtonomov, as well as 65th anniversary of Doctor of Agricultural Sciences V.A.Avtonomov. - Tashkent. 2006. C.34-36.

15. Avtonomov V.A. Geographically distant hybridization in selection of medium-fiber cotton varieties. - Tashkent: 2006. - C. 20.

16. Avtonomov V.A., Egamberdiev R. Inheritance of fiber yield in distant F1 hybrids of cotton G.barbadense L. //Mat.mezhd.nauchn.-practical conf. "State of cotton breeding and seed production and prospects of development", dedicated to the 110th anniversary of Academician A.I.Avtonomov, as well as the 65th anniversary of Doctor of Agricultural Sciences V.A.Avtonomov. - Tashkent. 2006. C.42-45.

17.Avtonomov Vik.A. Intraspecific geographically distant hybridization of cotton for resistance to wilt and black root rot. Dis.... of doctor of agricultural sciences in the form of scientific report.-Tashkent.2010,-78C.

18. Avtonomov Vad.A. Abstract of doctoral dissertation. 1993. C.40.

19.Aleksandrov A. Seed production of cotton // -Moscow. -1962. C.11-12, 125.

20.Alexandrov V.B. Darwinian selection on seeds, on their biological qualities // Selection and seed production. - Moscow.1937. - №8-9. - C.19-21.

21.Aleksandrov S.V. Seed production of greenhouse and greenhouse varieties of cucumbers and tomatoes. - Vegetable growing of protected ground. - Moscow: Selkhogiz, 1958. - 251 c.

22.Alekseev R.V. On the causes affecting the weight and size of tomato seeds and the relationship of these indicators with the productivity of the offspring // Proc. of Volgograd Experimental Station VIR. Volgograd Experimental Station of VIR. - Volgograd. 1973. -C. 129-132.

23.Alekseev R.V., Churkina L.V. Productivity of tomato seed progeny of different places of reproduction in the conditions of the Urals // Vegetable and melon crops. Issue 3 - Astrakhan, 1975. - C 194-195.

24. Alekhin A.G. Whether intrasort crosses in cotton seed production are justified // Cotton growing. 1966. №1. C.30-33.

25.Arkatova E.I., Malinina N.A. To the question of intrasort crosses// Cotton growing. 1966. №15. C.35-36.

26.Arkatova E.I. Production - high-quality, pure-variety seeds // Zh.: Cotton growing. - Tashkent, 1979. - №11. -C. 10-11.

27.Alibekov A., Novikov B. Seed selection // Zh.: Agriculture of Uzbekistan.

- Tashkent, 1972. - №2. - C.27-29.

28.Anikeev S.P. Climatic conditions // Science-based system of farming in Tashkent region of Uzbek SSR. - Tashkent, SAO VASHNIL, 1988. - C.4-11.

29.Aramov M.H. Scientific center on breeding and seed production of vegetable crops in the south of Uzbekistan // Main directions and prospects of breeding and seed production of vegetable, melon crops and potatoes: Theses of reports of international scientific-practical conference July 2-5, 2001. Tashkent-Termez, 2001. - C.3-8.

30.Asamov D.K. Beknazarov B.O. Influence of some physiologically active compounds on the initial phases of germination of cotton seeds of different storage period // Theoretical and practical bases and prospects of development of cotton breeding and seed production: Theses of reports of the International Scientific and Practical Conference - Tashkent, 2002. -C.106-108.

31.Abdullaev A.K. To the question about placement of different varieties of cotton on the territory of the Republic of Uzbekistan // Proc. of SANIGMI. - Tashkent, 1998. - Issue 158 (239). - C. 57-65.

32.Aidarov Sh.G. Concretization of choice of fruit branches and location of bolls on cotton bush in improving the quality of seed material // Materials of the International Scientific and Practical Conference. - Tashkent, 2006. - C.207-208.

33.Azzi J. Agricultural Ecology. Moscow: Foreign Literature, 1959. - C.354.

34. Alt V.V. Instrumentation and information support of agroindustrial complex / V.V.Alt // Information technologies, information-measuring systems in the study of agricultural processes: materials of the regional scientific conference "AGROINFO-2000" (Novosibirsk, October 26-27, 2000) / Russian Academy of Agricultural Sciences. Siberian branch. - Novosibirsk, 2000. - CH.1 - P. 32-40.

35. Balashova N.N., Balashova I.T., Shatilo V.I. Mechanisms of interaction "genotype - environment" in plants // Current state and prospects of development of breeding and seed production of vegetable crops. International Symposium August 9-12, 2005. Materials of reports, messages. Vol.2. - Moscow: VNIISSOK, 2005. - C. 43-77.

36.Baranov P.A. About the form-forming role of the environment // Yarovization. - Moscow. 1969. - №5-6. - C.35-37.

37. Berg R.L. Correlation Pleiades and Stabilizing Selection. Application of mathematical methods in biology. - L.: Izvo LSU, 1964. - Vol.3. - pp.23-60.

38. Bereznyakovskaya A.V. Combination of precocity and boll size in cotton hybrids. "Socialist agriculture of Uzbekistan". 1959. № 12. C.43-46.

39. Bessonova T.B. Drought-resistant early maturing variety of spring barley Dinat. / T.B.Bessonova, D.L.Kozub // Genetic aspects of breeding in Kyrgyzstan:

Collection of scientific works. - Frunze. 1986.

40. Bolshakov N.V. Improve the methodology and organization of work in primary seed production // J.: Breeding and Seed Production. Tashkent, 1986. - №1. C.34-37.

41.Bolshakov N.V. et al. Some features of seed production organization under frequent variety change // J.: Breeding and Seed Production. Tashkent, 1990. - №1. C.32-34.

42.Bocharnikova N.I. Recombinations in the genus LucopersiconTourn // Genetic bases of selection of agricultural plants. - Moscow VNIISSOK, 1995. - C. 76-81.

43.Brezhnev D.D. Influence of ecological and geographical conditions on the formation of properties and traits in plants. In book: Achievements in plant breeding. - Moscow: Selkhogiz, 1958. - C.112-116.

44.Bazhanova A.P. Seed selection within the bush of cotton G.barbadense, and its importance for increasing yield // ed. of the Academy of Sciences of Turkmen SSR, Ashgabat, 1960. - №5. - C.19-20.

45.Berdimurodov T. Kozubaev S.S., Zokirov S.T. Influence of the weight of 1000 pieces of cotton seeds on sowing and yielding qualities // Scientific bases of development of cotton and grain farming in farms: Materials of the international scientific-practical conference. - Tashkent, 2006. - C.467.

46.Buriev H.Ch., Zuev V.I., Kodirkhujaev O.K. Sabzavot ekinlari breeding, urugchiligi va urugshunosligidan amaliy mashgulotlar - Toshkent, Mekhnat, 1997. - C.84-93.

47.Butkevich V.V. // Methods and conditions of improved sowing mat6erial. M., VNIISSOK, 1959. - C.15-16.

48.Butkevich C.B. Yielding qualities of tomato and pepper seeds grown under different conditions // Proceedings of the Moldavian Research Institute of Irrigated Agriculture and Vegetable Growing. - Kishinev, 1965. Vol. VII, vol. 1. - P.112-115.

49.Vavilov N.I. Critical review of the current state of the genetic theory of plant and animal breeding // Genetics. 1965. №1. C.20-40.

50. Vavilov N.I. Genetics at the service of socialist agriculture. Ibid. 1966. M.: Kolos. C.262-287.

51.Varuntsyan I.S. On methods of production of elite cotton seeds (USA) // Zh.: Cotton growing. - Tashkent, 1971. - №№6,7. - C.12.

52.Varuntsyan I.S. About cotton seed production // Vesti s/kh nauki. - Moscow, - № 12 - P.87-90.

53.Vasiliev A.A. About preparation of bare cotton seeds for precision sowing

// Zh.: Cotton growing. - 1962. - №10. -C21-22.

54.Vahennym K.G. Influence of growing conditions of vegetable seeds and fodder root crops on breed qualities of varieties: Author's abstract. Diss. On coisk. Uch.st.st.kand.s/kh. sciences. - Tallinn: Estonian Agricultural Institute, 1955. - C.23.

55.Verkhoturtsev F.A., Khusanov R.H. Favorable zones of cotton sowing-. seed production // Collection of Urug sifatini oshirishning biologik va tekhnologik asoslari. - Tashkent, 1998. -C.19-20.

56.Verkhoturtsev F.A., Goldberg G.A. Fulfillment of seeds as an important selection criterion // Collection of Urug sifatini oshirishning biologik va tekhnologik asoslari. - Tashkent, 1998. -C.38-39.

57.Videnin K.F. Quality of seed grain and harvest // Selection and seed production. - Moscow, 1940. - №4. - C.20-21.

58. Gesos K., Ashirkulov A. Combination ability of varieties on fiber yield // Cotton growing. - Tashkent, 1986. №11. - C.29-30.

59.Gesos K.F. Nagymetov O. Character of inheritance of early maturity and productivity of ecologically-geographically distant hybrids and combinational ability of varieties. Tashkent, 1987, - p 65-71.

60. Gulyaev G.V. About methods and techniques of variety type preservation in primary seed production // Breeding and Seed Production. - 1970 - №6 - C.40-44.

61. Gulyaev G.V., Dubinin A.P. Heterosis and its use in plant breeding // Breeding and seed production. - Moscow: Agropromizdat, 1981. - C. 136-143.

62. Gulyaev G.V. Dictionary of terms on genetics, cytology, breeding, seed breeding and seed science. Moscow: Rosselkhozizizdat, 1983. - C. 240.

63.Goldberg G.A., Shpilevsky V.N. Seed selection by the mass of flyers // Zh.: Cotton growing. - Tashkent, 1983. - №9. C.36.

64.Goldberg G.A. Lomukhina G., Boltabaev H. Seed quality after sorting // Zh.: Cotton. - Tashkent, 1988. - №6. - C.37-38.

65. Goncharov P.L. Optimization of breeding process. / P.L.Goncharov // Improving the efficiency of breeding and seed production of agricultural plants: reports and reports. VIII genetic and breeding school (11-16 November 2001) / RAAS. Siberian branch of SibNIIRS. NSAU. - Novosibirsk. 2001. - C.5-16.

66. Guzhov Yu.L. Breeding and seed production of cultivated plants / Yu.L.Guzhov, A.Fuchs, P.Valicek. - M.: Mir, 2003. - C.536.

67. Darwin Ch. Origin of Species // Works Vol. 3. M.: Selkhozizdat. -1939э. - C.350.

68.Darwin C. The action of cross-pollination and self-pollination in the plant world. VOL. 6. -M. 1952. C.339.

69.Derevitsky V.A. About quantitative interrelations of different cotton varieties in Central Asian crops. //Cotton business. 1929. №3. C.282-287.

70. Derevitsky V.A. Novel data in the field of variation statistics. Supplementary chapters to the translation of Johansen's book, Moscow: Selkhozgiz, 1933.

71.Dobrutskaya E.G. Ecological bases of selection and adaptive seed production of vegetable crops. Diss. On coisk. Uch.st. doctor of agricultural sciences. -Moscow: VNIISSOK, 1997. - C.10-11.

72.Dobrutskaya E.G. Ecological substantiation - the basis of zonal placement of seed production // Potato and Vegetables. - Moscow, 2004. - №2. - C.11-13.

73.Dospekhov B.N. Methodology of field experience // M.:, - 1985 - P.351.

74. Janikulov F. Study of productivity and sustainability of cotton. //Mat. Interd.nauchn.-prak.conf. "State of cotton breeding and seed production and prospects of development", dedicated to the 110th anniversary of Academician A.I.Avtonomov, 80th anniversary of Academician S.M.Mirahmedov and Professor A.A.Avtonomov, as well as 65th anniversary of Doctor of Agricultural Sciences V.A.Avtonomov. - Tashkent. 2006. C.79-80.

75.Jafarov Sh. Influence of environmental conditions and harvesting dates of raw cotton on sowing qualities of seeds // J.: Cotton growing. - Tashkent, 1980. - № 11. - C.30-31.

76. Dragovtsev V.A. Algorithms of ecological and genetic inventory of gene pool and methods of designing varieties of agricultural plants by yield, stability and quality // - VIR, SPb.: 1994. - C.49.

77. Efimenko V.M. Increase of cotton fiber yield in the process of breeding and seed production. In V. Voprosy genetiki, breeding and seed production of cotton and alfalfa. - Tashkent: Nauka.1965. C.80-89.

78. Efimenko V.M. Cotton fiber yield. - Tashkent. 1976. C.109.

79. Yengalychev O.H. Maksudov E.Y. Combination ability of productivity, early maturity and inheritability of these traits in hybrids of ecologically distant varieties of cotton // Agricultural Biology 1985, №-4, -p 22-28.

80.Eremenko L.L. On the diversity of mature carrot seeds in terms of embryo size. // Botanical Journal, v.10. - Moscow, 1963. - № 48. - C.26-27.

81.Ergabulov J.E. Influence of intrasort crosses on variability of new varieties // Cotton growing. 1966. №4. C.27-28.

82.Zhalilov O.J., Adilov S. et al. Branching type and growth peculiarities of fruit branches in the lower and middle tiers // J.: Pakhtachilik va donchilik. - Tashkent, 1999. - №3. - C.4-6.

83.Zhuraev B.I., Abdalimov Sh. Seed quality and yield of cotton // Theoretical and practical bases and prospects of development of cotton breeding and seed production: Abstracts of reports of the International Scientific and Practical Conference - Tashkent, 2002. - C.102-103.

84.Zhuraev S.T., Namazov S.E., Muratov A. Combination ability of medium-fiber cotton varieties in terms of early maturity. - Tashkent. 2007. - C.233-235.

85.Zhuchenko A.A. Genetics of tomatoes. - Kishinev: Shtiyintsa, 1973. - C. 664.

86.Zhuchenko A.A. To the problems of scientific support of vegetable growing // Potato and Vegetables. - Moscow, 2002. - №2. - C. 3-6.

87. Zhuchenko A.A. Ecological genetics of cultivated plants (adaptation, recombinagenesis, agrobiocenosis) / - Kishinev: Shtiyintsa, 1980. - P. 587c.

88 . Zhuchenko A.A. Adaptive plant breeding (ecological and genetic bases). - Kishinev: Shtiyintsa, 1990. - C. 432.

89. Zhuchenko A.A. Adaptive potential of cultivated plants (ecological and genetic bases). - Kishinev: Shtiyintsa, 1988. - C. 767

90. Zaitsev G.S. Cotton. Botanical and agronomic sketch. - Moscow: Center.upr.printers of the All-Union National Economy of the former Soviet Union. 1925. C.54.

91. Zaitsev G.S. Cotton. - Tashkent. Л.1929. C.219.

92.Zaitsev G.S. Influence of temperature on the development of cotton // Proceedings of Turkestan breeding station - Issue 3. - M.-L.: Promizdat, 1930. - C.260.

93.Zakirov S.T. Seed quality management on the basis of ISO 9000 series standards // Materials of the International Scientific and Practical Conference. - Tashkent, 2006. - C.218.

94.Zakirov S.T., Kozubaev S.S. Improvement of cotton seed quality control system // Scientific bases of development of cotton and grain farming in farms: Mat. of international scientific and practical conference - Tashkent, 2006. -C.470.

95.Ibragimov Sh.I., Verkhoturtsev F.A., Kozubaev Sh.S. Restore the broken link between selection, variety testing and seed production // J.: Agriculture of Uzbekistan. -Tashkent, 1992. -№8-9. -C.3-5.

96.Ibragimov Sh.I., Verkhoturtsev F.A. Seed production of cotton - clear organization // Zh.:Agriculture of Uzbekistan. -Tashkent, 1992. - №4-5. -C.3-5.

97. Ibragimov P.Sh., Berdimuradov T., Kozubaev S.S. Yield qualities of cotton depending on the weight of 1000 pieces of seeds // Theoretical and practical bases and prospects for the development of breeding and cotton seed: Theses of doc. of interdisciplinary scientific and practical conference. - Tashkent, 2002. -

C.100-102.

98. Ivanov E.A. Modification variability of yield properties of wheat seeds under the influence of growing conditions.

99. Ignatov V.V. On the state and prospects of primary and elite seed production in northern Kazakhstan and Siberia // Bulletin of Seed Production in CIS, 1998. - №3. C.26-29.

100.Iksanov M.I. Guza breeding serviceblarning samaradorligi tugrisida // J.: Pakhtachilik va donchilik. -Tashkent, 2000. - №1. C.21-23.

101. Iksanov M.I. Egamberdiev A.E., Ibragimov P.Sh. Imminent issues of improvement of cotton seed production // J.: Pakhtachilik va Donchilik. -Tashkent, 2001. -C.5-7.

102. Iksanov M.I. Fiber quality of the newest varieties and lines of fine-fiber cotton. // Mat.mezhd.nauchn.-prak. Conf. "State of cotton breeding and seed production and prospects of development", dedicated to the 110th anniversary of Academician A.I.Avtonomov, 80th anniversary of Academician S.M.Mirahmedov and Professor A.A.Avtonomov, as well as 65th anniversary of Doctor of Agricultural Sciences V.A.Avtonomov.- Tashkent, 2006. C.192-194.

103. Iksanov M.I. Selection of long-fiber cotton varieties with the first fiber types. State and prospects of its development in Uzbekistan. // Mat.mezhd.nauchn.-prak. Conf. "State of cotton breeding and seed production and prospects of development", dedicated to the 110th anniversary of Academician A.I.Avtonomov, 80th anniversary of Academician S.M.Mirahmedov and Professor A.A.Avtonomov, as well as 65th anniversary of Doctor of Agricultural Sciences V.A.Avtonomov.- Tashkent, 2006. C.194-196.

104.Instructions for the production of elite and first reproduction seed for 1937, 1938, 1940, 1945, 1950, 1963, 1967, 1981.

105.Islamov I.K. To the question of the nature of fiber yield in cotton // Proceedings of the Tajik Agricultural Institute. Tajik Agricultural Institute. - Dushanbe, 1964. -C.48-61.

106.Yigitaliev M., Bekmuratova U. Reorganize cotton seed production // J.: Agriculture of Uzbekistan. -Tashkent, 1991. - №1. -C.4-5.

107.Imamaliev A. Scientific bases of high yields // J.: Cotton growing. -1981. - №6. -C.6-7.

108.Kazantsev I.A. Restore the former method of elite work // Cotton growing. 1966. №2. C.29-31.

109.Kanash S.S., Straumal B.P., Arutyunova L.G., Tribunsky A.N., Kratirov O.V., Tashlanov A.N. To the question of intra-variety crossing in cotton seed production // Cotton growing. 1965. №10. C.32-34.

110.Kakhkharov I.T. Correlation of early maturity trait with some economic-valuable indicators of medium-fiber cotton // Evolutionary and breeding aspects of early maturity and adaptability of cotton and other agricultural crops: Mater. International Conference-Tashkent: Fan Academy of Sciences of ReS.Uzbekistan, 2005.-P.110.

111. Kakhkharov I.T. Genetic bases of seed breeding in selection and production of pure-variety seeds of new cotton varieties // Mat. of the International Scientific and Practical Conference - Tashkent, 2006. -C.221.

112. Kirichenko V.E. Influence of ecological conditions on the quality of tomato seeds: Author's abstract. Diss. On coisk. Uch. St. Candidate of Agricultural Sciences. - Kharkov, 1977. - C.21.

113.Kiseleva A.G., Chernysheva S.P. Unjustified position // Cotton growing. 1966. №2. C.26-29.

114. Koloyarova L.F. Influence of cotton growing conditions on seed quality: Abstract of candidate dissertation. - Tashkent, 1958. -C.22.

115. Koloyarova L.F. Seed production of cotton in Uzbekistan. - Tashkent: Ministry of Agriculture of the Republic of Uzbekistan. 1962. C.59-61.

116.Koloyarova L.F. On the results of elite-seed breeding work with cotton variety 108-F // Issues of genetics, breeding and seed production of cotton and alfalfa. Issue 1. -Tashkent, 1965. -C.29-31.

117. Kokuev V.I. Genetics of cotton. Reference book on cotton growing. - Tashkent: SoyuzNIIKhI. 1937. C.22-39.

118. Konoplya K.S., Konoplya S.P., Gurbangeldiev S. Breeding for precocity and tolerance to unfavorable environmental factors. // Mat.mezhd.nauchn.-prak. Conf. 95 years since the birth of Acad. S.S.Sadykov. - Tashkent: FAN 2005. C.115-116.

119.Kratirov O.V., Akhmedov E.A. On the change of variety 108-F in the process of elite-seed work// Cotton growing, 1965. №4. C.21-23.

120.Kratirov O.V. Seed production of cotton // Reference book on cotton growing. - Tashkent, 1981. -C.129-130.

121.Kratirov O.V. Main causes and consequences of varietal purity reduction // Zh.:Cotton growing. -Tashkent, 1987. -№3. -C.7-12.

122.Kratirov O.V. Scientific approach or what represents a new method // Zh.: Cotton. -1988. - №4. -C.31-33.

123. Kurepin Yu, Imamaliev A. Criteria of variety homogeneity // Zh.:Cotton. -1988. - №4. C.40-43.

124.Kozubaev Sh.S. Order of multiplication of elite seeds in Uzbekistan and abroad // J.: Uzbekiston kishlok khuzhaligi. - Tashkent, 2002. - №2. -C.20-23.

125.Kozubaev S.S., Abdullaev F.A. Criteria of protection ability of new varieties of cotton // Theses of reports of the international scientific-practical conference. - Tashkent, 2002. -C.62-65.

126.Kozubaev Sh.S. Stimulation of sales of seeds and varieties // Theses of reports of the international scientific-practical conference. - Tashkent, 2002. -C.91-93.

127. Kozubaev S.S., Amanturdiev A.B., Shpilevsky V.N. Variety certification of seeds // Theses of reports of the international scientific-practical conference.- Tashkent, 2002 - P.93-96.

128.Kozubaev S.S. et al. Introduction of new methods of production of pre-elite and elite seeds. Way to preserve and improve varietal and yield qualities of multiplied cotton varieties // Theses of reports of the international scientific-practical conference. - Tashkent, 2002. -C.98-99.

129.Kozubaev Sh.S. et al. Low-flowered seeds and precision sowing of cotton // Theses of reports of the international scientific-practical conference. - Tashkent, 2002. -C.100-102.

130.Kozubaev S.S. et al. Yield qualities of cotton depending on the weight of 1000 seeds // Theses of reports of the international scientific-practical conference. - Tashkent, 2002. -C.109-110.

131.Kozubaev S.S., Nazarov R.S., Ibragimov P.S. Placement of cotton varieties in different agrocenoses of the Republic of Uzbekistan // Theses of reports of the Republican Conference Polymers-2002. - Tashkent, 2002. -C.8.

132.Kozubaev Sh.S. Scientifically-based placement of varieties - a real way to improve the quality and yield of cotton // Tashkent, 2003. -C.103-104.

133.Kozubaev S.S., Amanturdiev A.B., Zakirov S.T. Protection and certification of breeding varieties // J.: Uzbekiston kishlok khuzhaligi. - Tashkent, 2003. -№8. -C.19-20.

134.Kozubaev S.S. et al. New methodology of production of seeds of pre-elite and elite - as an important reserve of preservation of typicality of cotton variety at reduction of production cost // Theses of topics of the international scientific-practical conference. - Tashkent, 2003. -C.56-58.

135.Kozubaev Sh.S. Seed production: state and prospects. // J.: Uzbekiston kishlok khujaligi. - Tashkent, 2004. -№3. -C.11-12.

136. Kozubaev S.S. et al. Problems of seed production in Uzbekistan during the transition to the market. // Uzbekiston Bugdoy breeding, urugchiligi va etishtirish technologichesiga bagishlangan birinchi milli tuplami conference. - Tashkent, 2004. -C.64-66.

137.Kozubaev Sh.S. Varietal purity and seed renewal // J.: Uzbekiston kishlok khuzhaligi. - Tashkent, 2004. -№5. -C.17-18.

138.Kozubaev Sh.S. Guzaning original uruglari // Zh.: Uzbekiston kishlok hujaligi. - Tashkent, 2004. -№10. -C.10-11.

139.Kozubaev S.S. Main tasks of marketing of sowing seeds // Uzbekistan. Bugdoy breeding, Urugchiligi va etishtirish technologichesiga bagishlashgan birinchi milli tuplami conference. - Tashkent, 2004. -C.274-276.

140.Kozubaev Sh.S. Formation of agricultural seeds market // Bozor islohotlari ITI: Collection of theses - Tashkent, 2004. -C.30-32.

141.Kozubaev S.S. et al. Seed quality variability-objective reality // J.: Uzbekiston kishlok khuzhaligi. - Tashkent, 2005, -[1]1. -C.12-13.

142.Kozubaev S.S., Rashidova S.S. Capsulation as a reserve for improving seed quality // J.: Uzbekiston kishlok hujaligi. - Tashkent, 2005. -№10. -C.10-14.

143.Kozubaev Sh., Rashidova D., Rakipov V., Rafikov D. Seed diversity - objective reality. // Uzbekisto kishlok khujaligi. - Tashkent, 2005. - №1. - C.12-13.

144.Kozubaev S.S., Isaev R.S., Rashidova D.K., Shpilevsky V.N., Turabkhodjaeva M.. Influence of different mineral components in pre-sowing moistening of seeds on germination and yield of cotton // Scientific bases of development of cotton growing and grain farming in farms: Mater. inter. scientific and practical conf. - Tashkent, 2006. -C.477.

145. Kuziboev S.S., Mamarakhimov B.I. Monograph. Guza urugchiligini takomillashtirish omillari. - Tashkent, 2013. -C.22.

146.Kozubaev Sh.S.. B. Mamarakhimov, G. Abduvokhidov Improvement of methodology of elite-seed production of cotton. // Uzbekiston pakhtachiligini rivozhlantirish istikbollari. - Tashkent, 2014. - C.262-265.

147. Shukhrat Kozuboev, PhD, Bunyod Ikromovich Mamarakhimov and others. Variability of Agronomic Traits in Multiple Self-Pollinations. // Proceedings of the Tashkent International Innovation Forum. TIIF-2015. P. 289-293.

148.Kondratyeva I.Yu. Tomatoes for the middle strip of Russia. - Moscow. 2003. - C.115.

149.Konstantinov P.N. Influence of reproduction sites on seed yields and principles of seed supply to a variety plot. // Selection and seed production. - Moscow, 1956. - №5. - C.33-36.

150. Kravchuk V.Ya., Pivovarov V.F., Dobrutskaya E.G. Influence of different ecological and geographical conditions of vegetable seeds reproduction on their quality in the offspring. // Collection of scientific papers of VNIISSOK. - Moscow, 1987. - C.29-33.

151.Kravchuk V.Ya. Variability and inheritance of tomato yield depending on ecological zones of cultivation. // Collection of scientific papers of VNIISSOK. - Moscow, 1994. - C.17-19.

152.Krasochkin V.T. About zonality in vegetable crops seed production. // Materials of the meeting on exchange of experience in growing high yields of vegetable crops with the participation of representatives of the countries of people's democracy. - Moscow: Selkhogiz, 1958. - C.101-105.

153.Christidis S.B. Problems of cotton cultivation // M.: IL, 1959. -C.686.

154.Krug G. Ovoshchevodstvo. - Moscow: Kolos, 2000. - C.108-112.

155.Kariev A.A., Sultanova Y.M. et al. Varietal specificity of cotton responsiveness to mineral nutrition // Theoretical and practical bases and prospects of development of cotton breeding and seed production: Proc. of Interdisciplinary Scientific and Practical Conference. Tashkent, 2002. -C. 108-109.

156. Kim R.G., Khozhambergenov N. Cotton earliness and its interrelation with morpho-economic traits // Theoretical and practical bases and prospects of development of cotton breeding and seed production: Proc. of Interdisciplinary scientific and practical conference. Tashkent, 2002. -C.87-89.

157.Kim N.A. Good family - good tribe // Cotton growing. 1965. №12. C.34-35.

158. Kuptsov A.I. Elements of general plant breeding. Novosibirsk:, Nauka Sib.otd. 1971. - C.373.

159. Kovalev V.M. Theoretical bases of optimization of crop formation. - M.: TSKHA, 1997. - C.284.

160. Complex program "Semkhlopok" of the Uzbek SSR for 1988-1990. - Tashkent, 1988. -C.48.

161. Kushaliev A., Avtonomov V.A., Khalmanov B., Egamberdiev R., Normuradov D., Burieva M. Economic-valuable traits in selection of medium-fiber cotton. // Mat.mezhd.nauchn.-prak. Conf. "State of cotton breeding and seed production and prospects of development", dedicated to the 110th anniversary of Academician A.I.Avtonomov, 80th anniversary of Academician S.M.Mirahmedov and Professor A.A.Avtonomov, as well as 65th anniversary of Doctor of Agricultural Sciences V.A.Avtonomov.- Tashkent, 2006. C.97.

162.Kuchkarov S.K., Janikulov B., Saburov S.. Harvest qualities of melon seeds depending on the feeding area of parental plants, their cultivation. // Proceedings of the Research Institute of vegetable and melon crops and potatoes. Issue XI. Issues of breeding, seed production and agrotechnics of vegetable and melon crops and potatoes in Uzbekistan. - Tashkent, 1974. - C.43-48.

163.Lazarev A.V. Influence of agroecological conditions on seed productivity of Peking cabbage. // Potatoes and vegetables. -Moscow, 2004. - №7.

- C.31.

164. Larionov Y.S. Competition and intergenotypic relationships in spring wheat, their significance for breeding // Agricultural Biology. -1983. - №9. - C. 3-8.

165. Larionov Yu.S. Issues of seed production of grain crops (Theory and practice) / Yu.S.Larinov. - Kurgan, 1992. - C. 162.

166.Larionov Y.S. Methodology of evaluation of yield properties of grain seeds and its brief substantiation // Ways to improve the efficiency of agricultural production: collection of scientific articles. -Chelyabinsk, Chelyab. GAU, 1998. - C.69-76.

167.Larionov Y.S. Estimation of yield properties and yield potential of seeds of grain crops. -Chelyabinsk, Chelyab. GAU, 2000. - C.100.

168. Larionov Y.S. Theoretical bases of modern seed production and seed science. - Chelyabinsk: Chelyab. GAU, 2003. - C. 364.

169. Larionov Y.S. Management of variety adaptability / Y.S. Larionov, L.M. Larionova, E.P. Novokreshchinov - Chelyabinsk: Chelyab. GAU, 2004. - C. 301.

170. Leyshram D.K. Genetic analysis of morphological and economically valuable traits in interspecific hybridization of low-growing varieties of G.hirsutum L.and G.barbadense L.Avtoref. diss. candidate of S.Kh.N. Tashkent 1984.

171. Ludilov V.A. Seed production of vegetable and melon crops. - Moscow: Agroproimzdat, 1987. - C.15-26.

172. Ludilov V.A. Bring the seed production industry out of crisis. // Potato and Vegetables. - Moscow, 2004. - №2. - C.5-7.

173.Lukyanenko P.P. Selection by specific weight as a method of increasing yield qualities of seeds. // Selection and seed production. - Moscow, 1940 - № 3. - C. 16-18.

174.Lychko G.P. Productive qualities of seeds from different inflorescences in greenhouse varieties of tomato. // Collection of scientific works. - VNII of selection and seed production of vegetable crops. - Moscow, 1985. Vyp.21. - P.33-38.

175. Madrakhimov I.H., Hasanov H.E., Akhmedov J.H., Sharipov Sh.T. Study of sowing qualities of seeds of new and promising varieties of cotton // Theoretical and practical foundations and prospects for the development of breeding and seed production of cotton: Theses of reports of the International Scientific and Practical Conference - Tashkent, 2002. -C. 103.

176. Mazo E.Z. Save intra-variety crosses // Cotton growing. 1966. №1. C.33-35.

177. Makaro I.A. Increase of sowing qualities of seeds of vegetable crops. - Moscow, 1966. - C.121-123.

178 .Maletsky S.I. Epigenetic and synergistic forms of inheritance of reproductive traits in covered plants. Epigenetics of plants: collection of scientific works. - Novosibirsk: ICIG SB RAS, 2005. - C. 54-86.

179.Meredov Ya. Brief history and current state of cotton seed production in the USSR: Collection of scientific works. Ashgabat: Ylym, 1976. Vyp14. C.80-90.

180. Meredov Y., Meretkuliev B, Jumaev I. A new method of production of seeds of elite and first reproduction // J.: Cotton, 1988. - № 4.-C.29-31.

181.Mirjuraev M. Technological properties of fiber of cotton varieties and hybrids under different growing conditions. Author's thesis. Cand.diss. Tashkent. 1966.

182. Muminov K., Egamberdyev R., Mamarakhimov B. Difference of morpho-economic traits of seeds of variety C-6524 obtained from different elite-seed farms of the republic. // Agro Ilm 4[20] son, 2011.

183. Muminov K.H., Egamberdiev R.R., Mamarakhimov B.I., Shpilevsky V.N. The result of reproduction of seeds of elite variety Bukhara-6 obtained in different elite farms. // Zhakhon andozalaririga mos guza va beda navlarini yaratish istikbollari. Tashkent, 2012. C.336-339.

184.Muratov A., Namozov S.E. Relationship and role of fatty substances in the formation of fiber structure of fine and medium-fiber cotton varieties of different maturity. / / / Va boshka kishloq khujalik ўsimliklari evolutionary va breeding kirralari nomli halkaro ilmiy conf. materials.- Toshkent: Fan, 2005. - C.116-117.

185. Musaev D.A., Almatov A.S., Saidkarimov A.T., Bekmukhamedov A., Khayitova Sh. Inheritance of seed pubescence and fiber yield in lines of cotton genetic collection. // Mat.mezhd.nauchn.-prak. Conf. 95 years since the birth of Acad. S.S.Sadykov. - Tashkent: FAN 2005. C.106-108.

186.Mukhamedjanov M.V. On sorting cotton seeds by specific weight // Zh.:Cotton growing, 1964. - №8. -C.20-23.

187.Mukhamedjanov M.V., Soloviev V.P., Ibragimov Sh.I. High-quality seeds - the basis of high yields of cotton // Fundamentals of intensification of chemicalization of agricultural production in irrigated cotton growing: book by "Nauka" Publishing House.- Tashkent, 1965.-S.89-91.

188.Mukhin V.D. On the preparation of high-quality seed of vegetable crops. // Selection and seed production. - Moscow, 1977 - P.6.

189.Mukhin V.D. Characterization of planting and sowing material. Vegetable growing. - Moscow: Kolos, 2003. - C.103-112.

190. Mamarakhimov B.I. The state of modern seed production.// Akhborotnomasi Vestnik of Karakalpak branch of ANRUz. 2012. Nukus, No.2. C.36-39.

191. Mamarakhimov B.I., Shpilevsky V.N. et al. Computer program for rejection of elite material.// Zhakhon andozalaririga mos guza va beda navlarini yaratish istikballari. - Tashkent, 2012. -C.110-112.

192. Mamarakhimov B.I. Guzaning elita urugini etishtirishda yakka tanlov olish tartibi. // O'zbekiston Qishloq Xo'jaligi № 8. 2012. C. 15-16.

193. Mamarakhimov B.I., Kuziboev S.S. Elita urugchilik ishlarini olib borish serviceblari. // O'zbekiston Qishloq Xo'jaligi. No. 5 2012. C. 25.

194. Mamarakhimov B.I. Modal selection and selection on the complex of traits in the formation of new variety elite. - Tashkent, 2012. Agro Ilm. No. 2[22]. -C.8-9.

195. Mamarakhimov B.I. Guzaning birlamchi urugchiligi. // Khorazm Mamun Academyasi Akhborotnomasi .2012. C.20-23.

196. Mamarakhimov B.I. Guza navlarini zhoilashtirish va uning elite urughchiligi. // Khorazm Mamun Akademiyasi Akhborotnomasi. 2012. C.23-25.

197. Mamarakhimov B.I., Shpilevsky V.N. Monograph. Development and prospects of cotton seed production in Uzbekistan.// Tashkent. 2015.C.83.

198. Bunyod Ikromovich Mamarakhimov // Genetic Heterogeneity of Elite Materials of Commercial Varieties of Cotton in Nurseries. Proceedings of the Tashkent International Innovation Forum. TIIF-2015. P. 298-300.

199.Nazarov R.S. Cotton growing in the USA // Zh.: Cotton, 1992. -№3. - C.8-12.

200.Nazarov R.S. Cotton in Israeli // Zh.: Cotton, 1992. -№1. C.45-46.

201. Nazarov R.S., Ibragimov P., Kozubaev Sh. Placement of cotton varieties in different agrocenoses of the Republic of Uzbekistan // Polymerlar-2002: Resp. Anjum. Tuz. tup. - 2002. -C.8.

202. Nazarov R.S., Murtalibov M., Mamarakhimov B. State and prospects of cotton breeding in the Republic of Uzbekistan. // Uzbekistan Cotton and Textile Review. Tashkent, 2012. - C. 16-17.

203.Narimanov A.A., Kotov A.I. Sow bare seeds-no alternative // J.: Agriculture of Uzbekistan. -Tashkent, 1993, No.2. -C.4-6.

204.Narimanov A.A., Nugmanov R.Sh. et al. Varieties-Uzbekistan, fields-Kazakhstan, problems-common // J.: Agriculture of Uzbekistan. -Tashkent, 1993. -C.6-9.

205.Narimanov A.A. Guza urugchiligi muammolarini mazhmuaviy usulda hal kilish// Uzbekiston dehkonchilik sanoat mazhmuining ilmiy talimoti. -

Tashkent, 1995. - C. 437-442.

206. Narimanov A.A. Uruglarni certificatsiyalash yulida // J.: Pakhtachilik va donchilik. -Tashkent, 1997. -№3. -C.18-21.

207.Narimanov A.A., Uzakov Y.F. Scientific support of seed production // Urug sifatini oshirishning biologik va tekhnologik asoslari. -Tashkent, 1998. - C.17-18.

208.Narimanov A.A., Verkhoturtsev F.A. Urugulikni yukori unish energisi tula conli kuchatlar undirib olish garovi va uruglik uchun kulay mintakalar tanlab olishdir // J.: Pakhtachilik va donchilik. -Tashkent, 1998. -№4. -C.12-15.

209. Narimanov A.A. Mechanism of synthesis and replenishment of nutrients in seeds in interrelation with the mechanism of their germination. In Urug sifatini oshirishning biologik va tekhnologik asoslari. - Tashkent. 1998. C.37-38.

210. Narimanov A.A., Verkhoturtsev F.A. Primary seed production of cotton: problems, searches, solutions // Zh.: Agriculture of Uzbekistan. -Tashkent, 1999. - C.51-52.

211. Narimanov A.A. Monograph. 2000. C.146.

212. Narimanov A.A. High vitality of seeds as an important factor of their full-fledged germination // Fan. -Tashkent, 2000 -S.148.

213.Narimanov A.A. Scientific and methodological support of varietal renewal // Zh.: Agrarnaya nauka, Moscow, 2002. - №1. -C.20-21.

214. Narimanov A.A. High vitality of seeds. // Seeds. Moscow. 2002. № 13. C.24-26.

215.Narimanov A.A. Ways of introduction of market relations and economic efficiency in seed production // Materials of International Scientific and Practical Conference. - Tashkent, 2006. -C229-230.

216.Nenakhov I.F. Cotton seed production on an industrial basis // Cotton growing. -1979. -№ 7. -C.5-7.

217.Nikitenko G.F. Biological bases of seed production of grain crops, some issues of theory and practice // M.: 1980. -C.231.

218. Novolotsky V.D. Results of barley breeding using haplodia. Barley breeding and adaptability improvement to increase and stabilize yield: collection of scientific works. - Odessa: VSGI, - 1990.

219. Ovcharov K.E. Seed quality and plant productivity // Kolos, M.: 1966. - C.160.

220.Ovcharov K.E., Kizilova E.G. Variety of seeds and productivity of plants. - Moscow: Kolos, 1966 - P. 106-116.

221.Ovcharov K.E. How the cotton plant lives // Zh.:Uzbekistan. - Tashkent, 1974. - C.98.

222.Reports of the CCSHS for the years 1937-1991.yu

223.Pantileev Y., Smirnov I., Turbina V., Increase of field germination of seeds. // Potato and Vegetables . - Mos qua, 1971. - №12. - C.18-19.

224.Patron P.I. Intensive vegetable growing in Moldavia. Kishinev: Carta Moldavonescae, 1985. - C.177-179.

225.Pivovarov V.F., Dobrutskaya E.G. Use of ecological and geographical factor in selection of vegetable crops. // Collection of Scientific Works. - Moscow, VNIISSOK, 1987. - C.24.

226.Pivovarov V.F., Kravchuk V.Ya. Determination of adaptive capacity and stability of yield of new varieties in ecological testing. // Collection of scientific papers. - Moscow, VNIISSOK, 1988. - C.26.

227. Pivovarov V.F., Lebedeva A.T. Growing seeds on a homestead plot. - Moscow: Kolos, 1995. - C.78-79.

228. Pivovarov V.F., Mamedov M.I., Bocharnikova N.I. Paslenovye crops: tomato, pepper, eggplant, physalis. - Moscow, VNIISSOK, 1997. - C.30-70.

229.Pivovarov V.F., Balashova N.N., Balashova I.T. Development of priority directions of breeding and seed production of vegetable crops. // Reports of the scientific-practical conference. "State of problems and prospects of vegetable growing, melon growing and potato growing in the Republic of Uzbekistan". August 15, 2003, Tashkent, MAWR, 2003. - C.26-34.

230.Prokhorov I.A., Kryuchkov A.V., Komissarov V.L. Theoretical bases of seed production of vegetable crops. // Selection and seed production of vegetable crops. - Moscow: Kolos, 1997. - C.304-325.

231.Popov P.V., Daminova D.M. Conjugation of resistance to wilt and length of vegetation period on different infestation backgrounds. // Mat.mezhd.nauchn.-prak. Conf. 95 years since the birth of Acad. S.S.Sadykov. - Tashkent: FAN 2005. C.120-121.

232.Rakhimov H.R., Rudenko L.S. Seed science of cotton // Izd. Fan. - Tashkent, 1976. -C.167.

233. Rakhimov H.R., Kashkarova Z.Y. Narimov S. Seed production and seed science of cotton in Uzbekistan // -Tashkent, 1991. -C28.

234.Remidovsky Y.I. Cotton growing in the USA // -Tashkent, 1973. -C.67.

235.Rudenko L.S. Seed quality and productivity of offspring // Zh.: Cotton growing, 1973. - №10. -C.33-35.

236. Rashidova D.K., Mamarakhimov B.I., Shpilevsky V.N. Sowing with encapsulated seeds - a way to increase the seed multiplication factor. // Uzbekiston pakhtachiligini rivozhlantirish istikbollari. - Tashkent. 2014. -C. 274-278.

237.Rubtsov M.I.. Variety of vegetable plant seeds and its use in seed production. // Materials of the scientific conference on the problems of genetics, breeding and seed production of plants. - Gorki, 1969. -C.6-7.

238.Sadykov P.T. Character of variability of precocity and productivity of intervarietal hybrids of cotton depending on biological features and conditions of their cultivation. Avtoref. Diss. K.S.Kh.n. Tashkent, 1966. -24 C.

239. Saidaliev H. Gossypium avlodining tur hilma-hilligi va uning breeding-genetic izlanishlardagi ahmiyati. // Mat.mezhd.nauchn.-prak. Conf. "State of cotton breeding and seed production and prospects of development", dedicated to the 110th anniversary of Academician A.I.Avtonomov, 80th anniversary of Academician S.M.Mirahmedov and Professor A.A.Avtonomov, as well as 65th anniversary of Doctor of Agricultural Sciences V.A.Avtonomov.- Tashkent, 2006. C.19-22.

240. Saidkarimov A.T., Fork resistance of soon-to-ripen introgressive lines of cotton genetic collection. // Mat.mezhd.nauchn.-prak. Conf. 95 years since the birth of Acad. S.S.Sadykov. - Tashkent: FAN 2005. C.130-132.

241.Seyidaliev N.Ya. Primary seed production of cotton varieties // Theoretical and practical bases and prospects of development of cotton breeding and seed production: Tashkent, 2000. -C.89-90.

242.Semenchenko V.P. Change of economic traits depending on the place of reproduction // Zh.:Cotton growing, 1964. -№12. -C.31.

243.Sirota S.M. Organize the market of vegetable seeds. // Potato and Vegetables. - Moscow, 2005. - №1. - C.4-7.

244.Simongulyan N.G. On the genetics of early maturity of cotton // Cotton growing, 1968 № 2 -S.17-20.

245.Simongulyan N.G. Uzakov Y.F. Inheritance of elements of precocity and photoperiodic reaction in hybrids of annual forms with perennial form // Genetics, 1969, vol. b. p 24-31.

246.Simongulyan N.G. Problems of early maturity in cotton breeding. - Tashkent: 1971. - 221 C.

247. Simongulyan N.G. Problem of early maturity in cotton breeding. - Tashkent: FAN.1971. C.22.

248. Simongulyan N.G. Combination ability and inheritability of cotton traits. - Tashkent: FAN. UzSSR. 1977. C.140.

249.Simongulyan N.G. Genetics of quantitative traits of cotton. - Tashkent: Fan, 1991. -124 C.

250. Sprague D.F. Corn breeding. Corn and its improvement. M.: Izd-wo foreign. lit., 1957. - C. 163-222.

251. Stoletova E.A. Buckwheat - M.: Selkhozgiz, 1952. - C.179.

252.Straumal B.P. Breeding of early-ripening and high-yielding cotton varieties. // National Economy of Uzbekistan. 1961. - № 7, - C .34-36.

253. Strona I.G. . Divergence of seeds of field crops and its importance in shmenovodcheskiy practice. - Moscow: Kolos, 1964. - C.21-25.

254.Tadjiev K.M. Efficiency of bare cotton seeds in Surkhandarya Viloyat // Problems of development of cotton growing and grain production: Thes.doc. of interdisc. scientific-practical conf. -Tashkent, 2004. -C.110-112.

255.Ter-Avanesyan D.V. Role of insects in cross-pollination of cotton // Dokl. VASKHNIL. vol.1. - 1950. -C.26-30.

256.Ter-Avanesyan D.V. History of cotton growing// Cotton, M.: Kolos, 1973. C.5.

257. Ter-Avanesyan D.V. Cotton .// Cotton, M.: Kolos, 1973. C.483.

258.Tolmachev V.B. Efficiency of intrasort crosses // Cotton growing, 1966. №1. C.35-38.

259. Turabkhodjaeva M., Kozubaev S.S., Mamarakhimov B.I. et al. Study and improvement of crop seed standards. // Tashkent. Agro Ilm. C.19-20.

260. Terentyev P.V. Method of correlation pleiades. Vesti. Leningr. Un-ta .- 1959. - № 9.C.137-141; 191-240.

261. Umbetaev I. New varieties of cotton and the state of seed production in the Republic of Kazakhstan // Scientific bases of cotton and grain breeding in farms: Mat. of interd. scientific-practical conf. -Tashkent, 2006. -C.104-106.

262. Umbetaev, A.Kostakov, B.Mamarakhimov, Sh.Kozubaev/ On the use of modal selection for stabilization of the cotton species in the seed industry. Science and World. International Scientific Journal. №7 (23), 2015. P. 55-57.

263.Ushakova E.I. Tasks of selection seed-breeding work with vegetable crops. - Moscow: MUH, 1958. - C.187

264.Fursov V.N. Seed production of new varieties // Experimental mutagenesis and creation of initial breeding material of fine-fiber cotton. Ashgabat: Ylym, 1981. C.186.

265.Khoja-Ahmedov E.Y. Inheritance of some economically valuable traits. // Issues of genetics, breeding and seed production of cotton. - Tashkent, 1991. - C.3-6.

266. Shevelukha V.S. Plant growth and its regulation in ontogenesis. - M.: Kolos, 1992. - C.594.

267. Shuin K.A. Peculiarities of vegetable crops yield formation in different climatic zones. // Collection of scientific papers of the Gorki Agricultural Academy. - Gorki, 1967. - C.14-17.

268.Tsyba A. Cotton seed production to a proper level // Zh.:Cotton growing,

1983. -№2. -C.7-11.

269. Chub V.E. Natural-resource potential of the Republic of Uzbekistan. // Climate change and its impact on the natural-resource potential of the Republic of Uzbekistan // -Tashkent: Main Department of Meteorology under CM Roosevelt, 2000. -C.5-38.

270. Chulkina, V.A. Agrotechnical method of plant protection / V.A. Chulkina, E.Y. Toropova, Y.I. Chulkin, G.Y. Stetsov - M.: IVC "Marketing", Novosibirsk: LLC "Izd-vo YUKEA", 2000. - C. 336.

271.Egamberdiev A.E. Wild species of cotton - donors of fiber quality and wilt resistance. // DAN. UzSSR. 1979, № 8, c 66-57.

272.Egamberdiev A.E., Matchanov R. Cotton growing in Israel // J.Cotton. -Tashkent, 1991. -№3. -C.17-21.

273.Egamberdiev A.E. Development of breeding and seed production of cotton in the Republic of Uzbekistan // J.:Cotton growing. -Tashkent, 1993. -№4-6. -C.3-7.

274. Egamberdiev A.E., Ibragimov P.Sh., Ziyatov 3.3., Solihodjaev N.. Rayonized varieties in Uzbekistan // Tashkent, 1999. -C.38-40.

275.Egamberdiev A.E. Role of complex hybridization in improvement of selection-valuable traits of cotton // Theoretical and practical bases and prospects of development of cotton breeding and seed production: Thesis Dokl. -Tashkent, 2002. -C.89-90.

276.Yuryev V.Y., Kuchumov P.V. Michurin's ideas in the selection of field crops // Michurin's doctrine - at the service of the people. - Moscow: Selkhozizdat, 1955. Vyp.1. - C.11-15.

277. AVRDC - Asian Vegetable Research and Development Center // Vegetable production training manual. - Taipei: AVRDC, 1990. - Pub. № 90-328. - P.447.

278. Ahmed S.V., Saha H.K., Sharafuddin A.F.. Study of heterosis and correlation in tomato // Thai. J. Arg. Sc, 1988 - V.21. - No. 2. -P.117-123.

279. Buriev H.Ch. Status, problems and development perspectives of potato, vegetable and melon production in Uzbekistan // Vegetable Production in Central Asia. Status and Perspectives/ Worshop Proceedings June 13-14, 2003. -Amaty, Kazakhstan: AVRDC/ The World Vegetable Center, 2003. - P.117-128.

280.Carlos Augusto Avila, James Mac Stewart, Robert T. Robbins. Transfer of Reniform Nematode resistance from diploid cotton species to tetraploid cultivated cotton. The procceding of Cotton Beltwide conference, USA, 2005.

281 .Fang M.R., Mao R.C., Xie W.H.. Breeding cytoplasmicall mall sterile lines of eggplant. Actf Hortic. Sci. 1985. - P.12, 261.

282. Farkas J. Breeding for tomato fruit quality // Acta hortic. - Wageningen, 1988. - V.220. - P.69-75.

283. Khalil R.M., Midan A.A., Hatem A.K.. Breeding studies of some characters in tomato Lycopersicon esculentum Mill. - Acta Horticul, 1988. V. 220. - P. 77-83.

284 .Khapre P.R., Deokar A.B., Wanjari K.B.. Inheritanc of sptneness in Solanum melpngena x Solanum indicum. J. Maharashtra Agric. Univ. 1987. - P. 12,107.

285.Kruq H. Gemuseproduktion Ein Lehr-and Nachschlagewerk aur studibm and Praxis // Verlag Paul Parey. - Berlin and Hamburg. 1991. -P. 108-112.

286. Latterot, H. Les fusarioses de la tomate // Rev.hortic. 1989. - P. 29-32.

287. Madalageri B.B., Dharmatti P.R., Padanagur VG.M. Reaction of eggplant genotupes to Cercospora solani and Leucinodes orbonalis. Plant Pathol. Newsl, 1988. - P. 6,26.

288. Maiero M., No I.J., Barksdale T.H.. Inheritance of gollar rot resistance in the tomato breeding lines C 1943 and NOEBR-2. - Phytopath, 1990. - V.80. № 12. - P. 1365-1368.

289 .Manedov M.L., Pyshnaja O.N.. Breeding of sweet pepper for earliness. In: Eucarpia, IX- th meeting on Genetics and Breeding on Capsicum and Eggplant. - Budapest, 1995. - p. 120-124.

290. Pawar D.B., Mote U.N., Kale P.N., Ajri D.S. Promising resistant sourses for jassid and fruit borer in brinjal. - Curr. Res, Rep., Mahatma Arg. Univ., 1987. - P.3,81.

291. Rasmusson J. Effects of mass selection in mangest. Hereditas 1932. - № 16. - P. 249-256.

292. Perring T.M., Farrar C.A. Historical perspective and current world status of the tomato russet mite (Acari: Eriophidae) // Department of Enomol. Univ. of California Rev. - California. 1986. - P. 19.

293. Stroman G.H.. Variability and correlation in a cotton breeding program. "Jour.Agric.Res., 1954, p. 353-364.

TABLE OF CONTENTS

INTRODUCTION 4

CHAPTER i LITERATURE REVIEW 9

CHAPTER II. LOCATION, CONDITIONS, MATERIALS AND METHODOLOGY 35

RESEARCH ACTIVITIES 35

CHAPTER III. COMPARATIVE STUDY OF APPLIED METHODS OF ELITE SEED PRODUCTION WITH THE PURPOSE OF THEIR USE IN IMPROVEMENT OF COTTON SEED PRODUCTION SYSTEM IN THE REPUBLIC..42

CHAPTER IV. A COMPARATIVE STUDY OF THE RELIABILITY OF FAMILY EVALUATION IN SEED NURSERIES BY FIELD SCREENING AND LABORATORY ANALYSIS. 69

CHAPTER V. INBREEDING METHOD FOR SEED PRODUCTION (INDIVIDUAL SELECTION COMBINED WITH SELF-POLLINATION) IN PRE-BREEDING..112

CHAPTER VI VARIETY CERTIFICATION (IDENTIFICATION) OF SEEDS 133

CONCLUSION 149

REFERENCE LIST 152

Printed by Books on Demand GmbH, Norderstedt / Germany